Applications of Heat Pumps to Buildings

Applications of Heat Pumps to Buildings

Edited by A. F. C. Sherratt

Hutchinson

London   Melbourne   Sydney   Auckland   Johannesburg

This book is based on the second 'Heat Pumps for Buildings' conference organized in conjunction with the Chartered Institution of Building Services Engineers, the Institute of Refrigeration, the Royal Institution of Chartered Surveyors, the Department of Energy and the Department of the Environment.

Hutchinson Education

An imprint of Century Hutchinson Limited
Brookmount House, 62–65 Chandos Place, London WC2N 4NW

Longwood Publishing Group
27 South Main Street, Wolfeboro, New Hampshire 03894-2069

Century Hutchinson Australia Pty Ltd
16–22 Church Street, Hawthorn, Melbourne, Victoria 3122, Australia

Century Hutchinson New Zealand Ltd
32–34 View Road, PO Box 40–086, Glenfield, Auckland 10, New Zealand

Century Hutchinson South Africa (Pty) Ltd
PO Box 337, Bergvlei 2012, South Africa

First published 1987

Library of Congress Cataloging-in-Publication Data

Applications of Heat Pumps to Buildings.
    Based on the first conference organized in March 1984
    by the Construction Industry Conference Centre Ltd.
    Includes Index.
    1. Heat Pumps—Congresses. 2. Buildings—Heating
    and Ventilation—Congresses.
    I. Sherratt, A. F. C.
    TH7638.A67   1987   697'.07   87-4141
ISBN 0-09-161630-1

British Library Cataloguing in Publication Data

Applications of heat pumps to buildings.
    1. Heating.   2. Air conditioning   3. Heat pumps
    I. Sherratt, A. F. C.
    697'.07   TH7638
    ISBN 0-09-170750-1

The opinions expressed in the chapters are the contributors' and not necessarily those of the editor, the publisher, or the Construction Industry Conference Centre Limited

Typeset in 11 on 13pt VIP Plantin by
D P Media Limited, Hitchin, Hertfordshire

Printed and bound in Great Britain by
Butler & Tanner Ltd, Frome and London

Contents

List of Contributors

B. A. Bath, Partner, Rybka Smith & Ginsler, Toronto, Canada

T. R. Buick DPhil, Research Officer, Applied Energy Group, Cranfield Institute of Technology, Cranfield

R. R. Cohen BSc MSc MInstE, The Applied Energy Group, The School of Mechanical Engineering, Cranfield Institute of Technology, Cranfield

A. R. Coleman MA MSc CEng MInstE MInstR, Senior Engineer, Energy Division, W. S. Atkins & Partners, Epsom, Surrey

B. Dann CEng MIGasE MInstE, Technical Manager, Industrial and Commercial Sales, South Eastern Gas

K. J. V. Fowler FCIBSE FInstR M.ASHRAE, Principal Partner, Kenneth Fowler & Partners, Consulting Engineers, Wallington, Surrey

D. P. Gregory BSc PhD FCIBSE, Director, The Building Services Research and Information Association, Bracknell

G. Hamilton BSc MPhil, Senior Project Engineer, Systems Design & Performance Unit, The Building Services Research and Information Association, Bracknell

J. W. Kew MCIBSE, Head, Systems Design & Performance Unit, The Building Services Research and Information Unit, Bracknell

Dr-Eng Joachim Paul, Managing Director, Sabroe Kältetechnik, Flensburg, West Germany

J. Leary CEng MIMechE MCIBSE, Head of Environmental Engineering, The Electricity Council, London

B. S. Austin BSc, First Engineer, The Electricity Council, London

T. McDonnell BA CEng MInstE MCIBSE, Manager, National Accounts Department, Carrier Distribution Limited, Leatherhead

D. J. Martin BSc PhD MInstP, Project Officer, Energy Technology Support Unit, AERE, Harwell

P. W. O'Callaghan, The Applied Energy Group, The School of Mechanical Engineering, Cranfield Institute of Technology, Cranfield

D. R. Oughton MSc FCIBSE MInstR, Director, Oscar Faber Consulting Engineers, St Albans

S. D. Probert, The Applied Energy Group, The School of Mechanical Engineering, Cranfield Institute of Technology, Cranfield

J. P. Quick BA MSc, Principal Engineer/Analyst, Facet Limited

P. M. Robbins BSc, Assistant Surveyor, Northcroft Neighbour & Nicholson, Leamington Spa

V. Sharma BSc PhD CEng MIGasE MCIBSE, Head of Regional Activities Group, Central Laboratories, South Eastern Gas

R. B. Watson FRICS, Principal, Bruce Watson Associates, Coulsdon; *previously* Associate, Davis, Belfield & Everest, London

R. G. H. Watson BSc PhD CChem FRSC, Director, Building Research Establishment, Garston

Preface

The heat pump is a device which makes low temperature low grade heat useful by converting it to a higher temperature higher grade heat. There are prices to be paid for this conversion in terms of the mechanical work to effect the transfer and the capital cost of the sophisticated plant needed. The latter factor strongly determines the economics of installations alongside the cost of fuel to supply the mechanical work and the cost of an alternative supply of the heat pumped.

Present fuel price trends are not favourable to heat pump application but everyone knows that low oil prices can only be a short term phenomenon. A considerable number of heat pumps have been installed in UK buildings, even more in North America and the rest of Europe, in particular Germany. It is not yet a mass market but it is a developing one and market trends are examined at the end of this book alongside an examination and assessment of advanced heat pump technologies. Mistakes have been made and some of them are discussed in this book alongside successes.

This book deals specifically with application of heat pumps for building environment control. Air conditioning is increasingly popular in commercial buildings, indeed it is often installed in response to staff pressure even when other factors don't make it necessary. Any energy efficient air conditioning system must contain a heat pump – some installations need many small heat pumps – and this application is covered in some detail in this book. Heat pumps for heating only is however the main theme with chapters on system design, heat pump selection, heat sources and heat storage with a number of specific installations described in detail and a review of lessons to be learnt from experience in Germany. There is a particularly interesting case study of the heat pump installed in the *Edmonton Journal* building in Canada and case studies of several UK installations including the use of small heat pumps in the air conditioning system of a large office building.

Positive guidance on the design and installation of heat pumps in buildings is provided in this book. Each Chapter was presented at a major conference in London UK and the edited discussion provides a pier group assessment of the material presented.

The earlier book *Heat Pumps for Buildings* examines some fundamental issues and other projects and applications such as swimming pools, stores, hot water supply in small commercial buildings. This new book is a companion volume extending the scope of the consideration and the applications examined.

Heat Pumps systems are poised for rapid expansion when oil prices increase again. Clients building now may miss out by using economic evaluations of the moment. A wise client and consultant will seriously consider a heat pump system against realistic fuel price scenarios of the future.

A. F. C. Sherratt

Acknowledgements

The editor and organizing committee wish to thank all the people who have participated in the production of this book and in the arrangement and operation of the conference on which it has been based.

Special thanks are given to Ruth Brett, Diana Bell and their colleagues at the Construction Industry Conference Centre for their courtesy and efficiency in the organization and administration.

Organizing committee

Chairman: A. F. C. Sherratt BSc PhD CEng FIMechE FCIBSE FInstR

J. C. Barnes CEng MCIBSE MInstF
(representing the Chartered Institution of Building Services Engineers)

R. Gluckman MA CEng MIMechE MInstR
(representing the Institute of Refrigeration)

M. J. McCall BSc PhD
(representing the Department of the Environment)

D. J. Martin BSc PhD MInstP
(representing the Department of Energy)

D. Wilshire FRICS
(representing the Royal Institution of Chartered Surveyors)

1 Heat pumps: present practice, possibilities and promise

R. G. H. Watson

Introduction

At the Heat Pumps for Buildings Conference '83, Dr W. M. Currie of the Energy Technology Support Unit (ETSU) identified the following barriers to heat pumps [1].

1 High capital cost compared with conventional equipment.
2 Long pay-back periods at current fuel prices.
3 Doubts about reliability and performance.

At that conference several case studies were presented demonstrating the viability of the technology when correctly applied. Through publicized findings a clearer picture is emerging concerning heat pump performance, and there is greater confidence in the market concerning reliability than prevailed only a few years ago. However doubts still remain concerning the technology. Why? What problems remain and what of the future?

The installation of a heat pump involves entirely different hardware from that used in conventional heating systems. As with any emerging new technology, problems were experienced with the hardware but some installations proved unsatisfactory, not due to problems inherent in the heat pump itself but because insufficient attention was paid to the differences between a heat pump and its conventional counterparts. It is increasingly recognized that the problems of correctly sizing and integrating the heat pump, first with the heat distribution system, and then this in turn with the building, are major factors upon which either the success or failure of the heat pump ultimately rests. The questions of how this should be done and by what means remain issues for debate.

2 Applications of Heat Pumps to Buildings

Present markets

As heat pump operating times increase so do the running cost savings and hence the economic viability of any heat pump installation. The effect of run time on the payback period of an installation is illustrated in Figure 1.1, which has been constructed using the net present value method common within Government Departments as it encompasses a 'real rate of return' on investment. Prime building targets for the incorporation of heat pump systems are therefore those with a continuous and preferably constant heat demand over the full 24-hour period.

The most attractive example in this category is the swimming pool installation where energy consumption is dominated by the need to dehumidify the pool hall, to prevent condensation on and within the building fabric. As shown in Figure 1.2, the ability of a heat pump to dehumidify air at the evaporator and provide useful heating at the condenser enhances the heat pump's effectiveness, resulting in very favourable economic arguments for its installation. The outcome has been a proliferation of such installations, the majority of which are

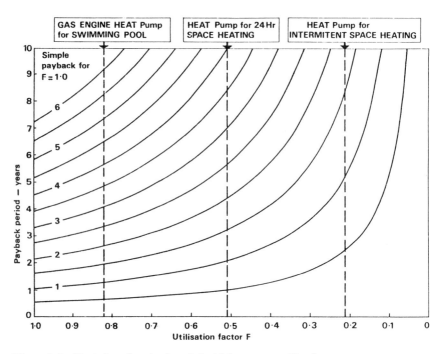

Figure 1.1 *Variation of payback period with heat pump utilization.*

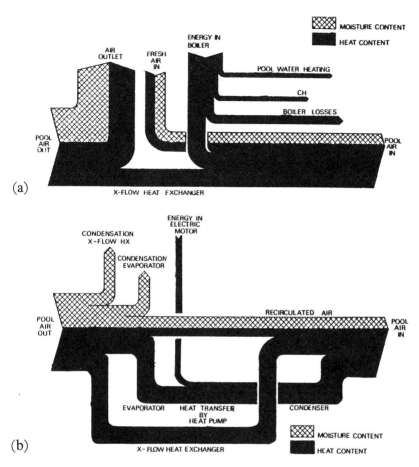

Figure 1.2   (a) Conventional boiler system with X-flow heat exchanger. (b) Electrically driven heat pump with X-flow heat exchanger.

electrically driven. An increasing number of gas engine-driven heat pumps which offer possibilities for heat recovery from burnt gases are entering the market [2, 3] (Figures 1.3 and 1.4).

Equally favourable opportunities for heating-only heat pump applications have been difficult to identify and the next best options include hospitals and old peoples homes. The EEC have recently provided financial assistance for the design of a new heat pump installation at St Mary's Hospital, Isle of Wight, and ETSU, acting on behalf of the Department of Energy, have recently commissioned field trials at two old peoples homes, one at Longlands Oxfordshire and the other at a site

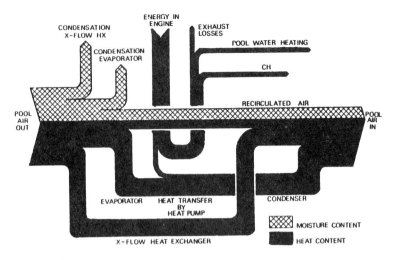

Figure 1.3   *Gas engine HP and X-flow heat exchanger.*

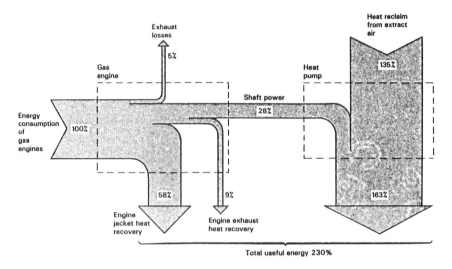

Figure 1.4   *Energy efficiency of a gas engine heat reclaim heat pump system.*

in Cheshire. These sites will provide important information on financial aspects of the installations' viability, test the reliability of the technology and, if successful, stimulate their promotion.

A considerable share of the heat pump market is taken up by machines which use air as the distribution medium for the heat pump's output. Units of this type range in size from 4 kW up to 100 kW output or more. Packaged units are the norm, reflecting trends in the air conditioning trade for cooling plant, and consequently are less expensive than individually designed units. They have found diverse application in the market in a multitude of outlets, including commercial premises, shops, factory units and hotels.

The cost and risk

The majority of heat pump installations are more costly to install than conventional alternatives. This means that the economic arguments are centred around reduced running costs for the heat pump, and analysis must include maintenance costs. Authenticated cost information of this nature is difficult to obtain and maintenance costs remain an unknown spectre frequently dependent upon the reliability of unproven plant. Such information is however a crucial prerequisite for the consultant to win support for a proposed heat pump installation.

A lack of information on the financial benefits, if any, to be derived from a heat pump installation feeds suspicion of their viability. Heat pumps will not increase their market share substantially until more favourable financial information is available.

Despite the number of heat pump installations now commissioned in Britain, applications are diverse and clients often find it difficult to identify an equivalent to their own proposals for which financial expectations have been confirmed. In this respect the British heat pump industry is still in its infancy. To the client, the heat pump is still a new and risky technology.

The future

The future of heat pumps is thus dependent upon improving:

1   The reliability of the plant.
2   The economics of the installation.

Should the reliability and economics be assured there remain two other impediments to the deployment of heat pumps:

1   Reluctance of customers to endorse 'new' technology and their prejudices on heating methods.
2   Difficulties of site installation.

The first is largely the result of unfamiliarity and will be overcome by the good publicity that should attend properly installed and economic heat pump systems. Where retrofit installations are involved (representing up to two-thirds of the conventional heat pump market [1]) prejudice may be related to concern over the effective utilization of existing space by the current methods and concern for other factors such as noise. These also relate to the second impediment and will be overcome by the growth in flexibility such as will be possible with low temperature methods of air distribution and underfloor heating systems.

Improved reliability

As heat pump designers, systems engineers and installers gain experience, so the reliability of installations will improve. Evaluation of low cost design, from lessons learned through mistakes and the discovery of unexpected problems when commissioning installations, seems likely to be more fruitful than high cost design choices *ab initio*. Dissemination of information relating to problems will be inhibited by commercial interests, but the growth of heat pump customers does suggest that the problem of reliability is being reduced to a tolerable level.

Reduced costs

There are a number of avenues through which the economic case for a heat pump can be enhanced. All of these reflect themselves either in the capital cost of the heat pump and its installation, or the running cost savings. If we examine the former and consider the heat pump unit itself, then it is unlikely that present costs can be reduced through technological advance. Cost reduction through volume production of packaged units is more likely. The majority of existing heat pump installations have been individually designed, reflecting the specific needs of the buildings concerned. An alternative route, which has been applied successfully to installations in supermarkets, is the deployment of a number of small packaged units which meet the heat demand of the building as a whole. Such systems have proved cheaper to install than single large machines tailored to the maximum requirements of the building. There are additional advantages in terms of capacity modulation, which has proved a problem on large single machines, enabling the best possible performance under a wide range of operating conditions.

The importance of considering the total integrated system when considering a heat pump has been demonstrated in studies undertaken by BRE and the Electricity Council, particularly with complex interactive systems such as exist in the application of heat pumps to swimming pools. The simplest system might appear to be the least expensive option but further analysis of more complex systems can result in a revision of necessary plant capacity and lead to installation cost reductions.

Improvements in running cost savings

Since increased utilization of a heat pump results in increased savings, it is common practice to install a heat pump alongside auxiliary conventional heating equipment, used when the heat pump's output is insufficient to meet the building's needs – the so called bivalent system. Optimization of a bivalent system under widely varying conditions, to maximize heat pump utilization factors while simultaneously minimizing fuel consumption, requires complex control. Such control equipment can often account for more than 20% of the installation costs. In the past the failure of such systems to perform efficiently has led to excessive and unnecessary run times, inefficient operation of heat pumps and unnecessary operation of auxiliary heating. The result is an installation which fails to realize the evisaged running cost savings.

Highly interactive machinery, including refrigeration equipment, engine drives, alternators, air dampers and other flow control devices are generally controlled with a variety of individual proportional band controllers and timers, which can be off-calibration, and function alongside, but isolated from, separate controller units which are specific to certain items of equipment. As a result, installations have been commissioned only to find the control system and logic inappropriate, or rendered so by differences between the envisaged and realized operating characteristics of the total system. Controls are 'hard wired', presenting little opportunity for corrective adjustments after installation without incurring heavy costs. There is scope here, with the growing trend towards the use of computerized energy management systems, for the use of software control of both heat pump and system operation. This would enable modifications to the control logic to be made after commissioning, with the minimum of effort helping to maintain optimum plant efficiency. An equally important improvement will relate to the ability for the software to include elements of system control that take into account complex considerations of the interactive nature of plant items among themselves and with the building.

Running costs, and hence running-cost savings, are sensitive to the thermodynamic operating co-ordinates of the refrigerant cycle. Performance and COP are improved as source and sink temperatures approach one another, which means high source and low sink temperatures, but this is as yet little exploited. The habitual choice of source is the ubiquitous atmosphere, freely available but not particularly suitable. An exhaustive study by W. S. Atkins [4] endorsed the practicality of ground water as a source, particularly for installations above 100 kW output. As yet few such installations have proceeded in the UK. A likely explanation is that the added complication, and often cost, of a ground water source is too much to contemplate. Similarly the use of building foundations or cellar walls and floors as a stable, relatively high temperature, source has not been exploited. It would appear that the air source heat pump will first have to demonstrate itself through a growing number of successful installations before more advantageous but initially more costly source applications will find willing sponsors.

The problem of tolerating lower sink temperatures for the benefits of COP are exacerbated by the fact that current upper limits on sink temperatures, available on vapour compression machines, are already low compared to those required in conventional wet radiator systems. This can be overcome, in modern premises especially, through the use of more extensive or more efficient radiators. This measure is however detrimental to the economic case for installation. Retrofit installations can profit from a total package approach incorporating improvements in insulation standards and air tightness for the building, measures which ameliorate radiator sizing problems. Improvements in building insulation standards lead to a reduced heat pump capacity and perturb the economics of installation.

An obvious possibility to exploit the advantages of heat pump operation at lower sink temperatures is air distribution. Here again, intervention with building design to allow a large flow of warm air without introducing draughts and other unacceptable effects could result in reduced heat pump costs. A further possibility is underfloor heating, as yet a fringe activity in the UK, but common in Europe, particularly West Germany. The heat pump is very suitable for use with underfloor heating systems but, since they are little used in the UK, suitable only for purpose-built buildings. A 40 kW installation has recently been commissioned for the Mitchell Museum in Southampton (Figure 1.5) at a cost deemed to be within that for a conventional system. This particular site also utilizes ground water as a source, using two bore holes.

Figure 1.5 *Mitchell Hall of Aviation, Southampton: water-source heat pump and underfloor heating.*

Possible future trends and developments

The gas engine-driven heat pump which has recently begun to prove its worth through a small number of pilot installations in the UK [5, 6] and many abroad, is likely to find increasing application in the retrofit market, where the provision of high temperatures remains an important requirement. Heat from burnt gases recovered by the engine cooling jacket and exhaust gas heat exchangers can be used to boost the heat pump's output temperature to about 70°C or provide separate water heating up to 90°C. The energy schematic of a typical gas engine heat pump is shown in Figure 1.3. A pilot scheme to be launched shortly through joint DTI/PSA sponsorship for heating a PSA regional headquarters office block at Manchester will be monitored by the BRE (Figure 1.6). A limited amount of monitoring equipment is already in place.

As improvements in reliability are made the natural development for

Figure 1.6   *Ashburner House, Manchester, PSA regional HQ building: gas engine-driven heat pump being installed.*

gas engine driven machines is the provision of combined heat and power (CHP). The concept is to increase the utilization factor of the heat pump's engine drive by powering an electrical generator, either simultaneously, or during heat pump off periods. Although some proposed installations will not be economically viable due to adjustments in fuel tariffs for CHP use, the improved attractiveness of CHP will inevitably widen the potential market. A number of installations now exist in the UK such as St Mary's Hospital, Isle of Wight and an ETSU-assisted project at Wallingford, which incorporates a gas engine and a 600 kW alternator. Regner [7] has recently published performance figures for a gas engine-driven CHP installation which can deliver 690 kW heating power and provides 100 kW electrical energy (Figure 1.7).

Installations which provide warm air heating are also likely to benefit from a broadening of their application to provide cooling as an extra. Cooling provision is seldom considered a necessity in the UK, but air distribution systems for heating are frequently adopted when environmental conditions such as noise or pollution problems become appreciable. Once a heat pump for warm air heating is considered, then the possibility of providing cooling for short periods of the year when the UK climate warrants its use, becomes an optional extra, available for little additional capital cost. Cooling provision substantially improves the status of a building, a valuable asset for high rent office stock and prestigeous premises for high technology companies. It is very probable that these factors will result in an increasing number of such installations.

At the 1983 Heat Pumps for Buildings conference the enhancement of economic arguments in favour of heat pumps when both heating and cooling could be simultaneously employed was identified [8]. To follow this up an ETSU demonstration scheme was commissioned, which involved the field trial of some ten installations in public houses to provide cellar cooling and hot water for washing purposes. A favourable report on the scheme has recently been released by ETSU [9] and is likely to promote further applications.

Since it will be many years before thermal stations cease to be the main source of electricity, it seems surprising that this energy-intensive power supply is used in the great majority of UK heat pump installations, instead of a direct thermal system. The absorption cycle heat pump has, however, received significant attention abroad, especially Japan [10] where a number of machines with power outputs in excess of

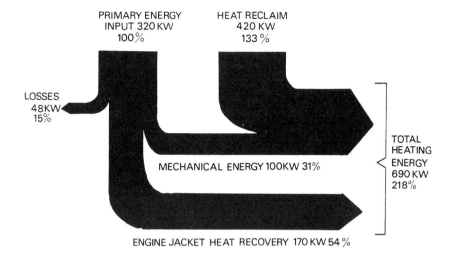

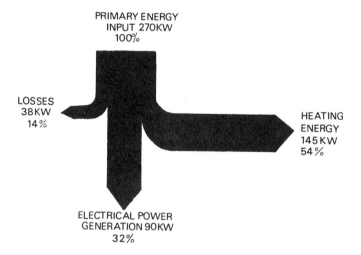

Figure 1.7   *Operating modes of CHP plant – Mannheim.*

100 kW have been successfully installed. Within the UK the absorption machine's prospects are generally viewed with scepticism and there are no large scale installations. The problems of low COP and undesirable use of corrosive chemicals and high pressures is frequently mentioned. Work is proceeding to develop new and suitable fluids for absorption machines [11, 12]. In Germany and elsewhere in Europe however, absorption heat pumps operating with established fluids have

found successful application. Absorption machines have been used for office heating with heat requirements in excess of 1 MW. Examples include the Siemens building, Munich, and a 1.7 MW installation at a Federal administrative complex in Meckenhiem [13]. A potential for absorption cycle machines in the UK has been identified by the DEn which is funding a significant number of collaborative research and development projects within UK industry with a view to eventual commercial application. The EC is also funding the UK development of a 300 kW absorption machine using lithium bromide and water as fluid combination [14]. Completion of this work is expected around 1990.

The possibility of using energy façades on buildings to collect energy from a number of climatic features, such as solar radiation, naturally convected air, rain, etc. is also receiving some attention abroad. These employ glycol/water mixtures or direct use of evaporative refrigerants for the circulating medium. The Technical University of Munich has assisted with the design of a 0.5 MW installation for a medical centre in Regensburg [15] incorporating these features, which are largely unproven and experimental at present.

Work on ground and groundwater heat pumps has led to proposals to embed heat collection pipes in the foundations of buildings which then act as a heat store and provide stable environmental conditions in the immediate vicinity of the pipes, frequently a problem with soil source heat pumps [16]. EC support is currently sought for an installation of this kind for offices of the Wilton Fijenoord shipping company in the Netherlands. Of concern with such projects is the possible damaging effects of pipe fracture or ground freezing, an effect which could cause cracking of the building structure [17] due to the expansion caused when water freezes.

Conclusion

No miraculous advance has appeared on the horizon to transform the heat pump industry. A gradual broadening of markets through a growing awareness of the heat pump, its potential and ever improving reliability is a realistic assessment of the medium-term prospects for heat pumps. This can, and is, being achieved by the steady appearance of pilot schemes, which demonstrate the viability of the technology, provide useful financial information and help promote further replications. More financial information on existing installations could rein-force this process.

Current installations must secure market acceptance before more adventurous, and ultimately more efficient, heat pump systems will be able to enter the market. For example, groundwater source heat pumps, intrinsically more suitable and efficient than air-source machines, deserve greater prominence in the market place.

It is clear from experience abroad that there is a role to be played by the absorption cycle heat pump for commercial premises, as yet lacking application in the UK, although research and development is proceed-ing towards the production of units for commercial application.

A growing realization of the importance of total system design and a thorough appraisal of system requirements is contributing to improve-ments in current system design, and ultimately system reliability.

In short, the future of heat pumps for buildings looks secure but destined to a protracted evolution before their full potential is realized.

The views expressed are those of the author, not necessarily those of the Establishment or the Department of the Environment.

References

1   Currie W. M. & Martin D. J. (1983) *Building energy conservation – the role for heat pumps*. Proceedings of the Heat Pumps for Buildings Conference, April 1983.
2   Weller J. W. (1983) *The gas engine dehumidification system at Farnborough Swimming pool*. Proceedings of the Heat Pumps for Buildings Conference, April 1983, Paper 8A.
3   McCall M. J. (1983) *Monitoring of the Farnborough Swimming pool system by the Building Research Establishment*. Proceedings of the Heat Pumps for Buildings Conference, April 1983, Paper B.
4   Report for the DEn by W. S. Atkins & Partners *Ground water source heat pumps for space heating in the UK*. Available from W. S. Atkins & Partners, Ashley Road, Epsom, Surrey.
5   See 2 above.
6   See 3 above.
7   Regner R. (1984) 'The combination technique of gas heat pump/block heat and power plant – tandem heat/power heat pump THKW'. *Fernwarme international* May (3) 1984. Source language of paper, German.
8   Ure J. W. (1983) *Small air-source heat pumps for hot water supply in commercial premises*. Proceedings of the Heat Pumps for Buildings Conference, April 1983.
9   ECDPS Report No. 117 *Heat pumps for cellar cooling and water heating in licensed premises*. Available from DEn through ETSU of the AERE, Harwell.
10 Watanabe H. (1984) *Field experiences with large absorption heat pumps (type 1) and*

heat transformers (type 2). Proceedings of IEA Heat Pump Conference, Graz, May 1984.

11   Bokelman H., Ehmke H. J., Renz M., Steimle P. (1983) *Working fluids for sorption heat pumps*. Commission of the European Communities' Proceedings – International Seminar 'Energy Savings in Buildings', Hague 1983, Paper A.1.18.

12   Cheron J., Durandet J. (1982) *Absorption–resorption heat pump. Investigation of solute–solvent pairs*. Commission of the European Communities' Proceedings of four contractors meetings on 'Heat Pumps', Brussels, 1982, Paper 4.3, pp. 312–327.

13   Adler H. 'A large scale absorption heat pump installation with earth collectors' *Haustechnische Rundschau*: 1081 v80(3), 201–204 (German only). English translation available from Building Research Establishment UK.

14   Heppenstall T. *A theoretical and experimental investigation into the performance of absorption cycle heat pumps applied to industrial processes*.

15   EC application, EE 287/83/D.

16   See 4 above.

17   Croney D. & Jacobs J. C. *The frost susceptibility of soils and road materials*. TRRL Report 90.

Discussion

D. Guy (Denis Guy & Partners)   In Canada, heat pumps are used to heat domestic buildings in the winter and cool them in summer. Are there many houses in this country in which a heat pump is used and if so with what success?

A. R. Coleman (W. S. Atkins & Partners)   There is a limited number of domestic heat pump installations in the UK at present. The heating duty for normal houses is generally too small for a heat pump installation to be economic, particularly when natural gas is the alternative heat supply. The cost of heat delivered by the heat pump during the day approaches the cost of that delivered by natural gas. There is then very little running-cost saving to off-set the substantial first cost of the heat pump system.

Dr R. G. H. Watson (Building Research Establishment)   Another factor must be the relative merit of investing money in better insulation rather than better efficiency in the use of energy. Typically, spending £100 on insulation will probably result in a better return than spending £100 on a heat pump-like system.

J. Torrance (Steensen Varming Mulcahy & Partners)   There is a significant feeling in the industry voiced at a CIBSE Scottish Region

Conference on heat pumps in 1983 that there tends to be an over-sale of unitary heat pumps for the domestic market generated on a 'sales' basis rather than on a properly considered engineering systems basis; perhaps there are more installed than we think but not as many performing as well as they ought.

D. R. Oughton (Oscar Faber Consulting Engineers)   There are a number of operating problems with domestic heat pumps. One is the availability of a suitable power supply. Dependent upon the size of the heat pump unit there may be a requirement for a three-phase supply which is not normally available to domestic accommodation from the Electricity Supply Boards. Use of several smaller heat pumps can perhaps get around this. Noise is also a potential problem particularly where dwellings are situated close together.

There *is* a market for heat pump installations in domestic accommodation located in remote parts of the country where a supply of natural gas is not available and is not likely to become available in the foreseeable future.

A. R. Coleman (W. S. Atkins & Partners)   Specific site circumstances can favour a heat pump. The owner of a large country house decided to install an ornamental lake. The possibility existed to use the necessary borehole to provide heat as well as water for the lake. Unfortunately, as Mr Oughton indicated, the cost of providing a three-phase electrical supply proved too high.

G. W. Aylott (The Electricity Council)   The number of heat pumps installed in domestic premises runs into four figures and yet finding 24 domestic heat pump installations on which to conduct trials is a phenomenal problem. It has indeed been necessary to conduct the trial with 16 installations. Direct use of off-peak electricity with water storage heating systems may in fact be much more cost-effective than domestic heat pumps.

Dr M. A. Bell (Plymouth Polytechnic)   With the increasing move towards better thermal insulation and draught protection standards for buildings, does this result in an advantage for the installation of a heat pump or a disadvantage? In other words, does it render the installation of a heat pump a better or a worse proposition? Dr Watson mentioned that one way of improving the economics of a heat pump is perhaps to adopt floor heating systems. These are much more widely adopted in Europe than in the UK. What are the reasons for this?

Dr R. G. H. Watson (Building Research Establishment) With improvement in the thermal performance of buildings, particularly houses, the scope for dramatic savings in energy costs will be reduced, and so the advantage that the heat pump brings in reduction of running costs with no advantage in capital cost is likely to decrease, and the opportunities for sales of heat pumps for that purpose will tend to go down rather than up if the trend towards better thermal insulation develops. Fuel prices are also important, while they have been steady or even falling relative to general inflation; around the year 2000 fuel prices will rise again, with increased heat pump sales. Elsewhere in Europe the combination of higher fuel or energy cost and the greater attention to living standards and comfort have provided better opportunities for heat pumps to be more effectively exploited in a better market. I would like to see more attention paid to choice and the systems approach to make the full use of all the apparently marginal but perhaps critical issues that the heat pump can offer.

R. Gluckman (W. S. Atkins & Partners) It is clear that the economics of heat pumps is marginal in many cases and improvements can be made by attention to heat sources and equipment selection. An important area often ignored is compressor selection. The isentropic efficiency of refrigeration compressors vary considerably.* Machines on the market have efficiencies between 40 and 80%. This in turn leads to a 100% variation in running costs. Bearing in mind that economics are marginal, I suggest that compressor efficiency must always be taken into account.

Small packaged units were mentioned in this chapter. This brings us to a dilemma. This type of package uses semi-hermetic compressors which are cheap and reliable but inefficient. Could Dr Watson comment on the use of such compressors; should we be using better 'industrial' compressors with high efficiency?

Dr M. McCall (Building Research Establishment) I would completely agree with Mr Gluckman. There can be a trade off between compressor efficiency and system efficiency. For example, an advantage of multiple rooftop units is that capacity modulation is available by switching on only the number of units needed to exactly satisfy the load, whereas a single large unit would suffer from

---

* *Industrial Applications for Heat Pumps* Conference September 1984, York. BHRA Fluid Engineering, Bedford 1984.

performance degradation because of cycling effects which would counter the possibly higher efficiency of such a large unit. Predictability is also important. A packaged unit will have had quite a lot of research effort put into it to optimize its performance, whereas single larger machines are generally one-off, purpose-built designs and the scope for error is very much greater.

J. Nisbet (J. Nisbet & Partners)   What is the effect of heat pumps on the design of buildings? Do they require more space and would they alter the shape and size of buildings? Rooftop units can for example look less like a carbuncle and more like a Second World War anti-tank trap. Is the use of heat pumps going to have an adverse effect on the design of buildings?

Dr R. G. H. Watson (Building Research Establishment)   The general answer is 'no'. What seems to be important is that the kind of questions being addressed as a result of interest in heat pumps will have a valuable influence on the designers of buildings and the dialogue they will have with their heating and ventilating colleagues in optimizing building design. Therefore, although building size will not be affected, heat pumps may result in optimizing the design of both the building and its mechanical services system.

M. D. Terry (North Thames Gas)   Dr Watson suggested a possible decrease in the application of heat pumps in future low-energy housing. Looking at new stock of such houses with high insulation and low ventilation rates there is a case for very small heat pumps to act as dehumidifiers. A small through-the-wall type of unit is already on the market.

Dr R. G. H. Watson (Building Research Establishment)   BRE Scottish Laboratory is reviewing the use of dehumidifiers. Social aspects are a major consideration. People with condensation problems generally cannot afford either the energy or the machinery to get rid of the water. The house is closed up to save heat, fills with steam, and the result is condensation. Telling these people that a heat pump saves energy is unhelpful because they are not able to afford one. Perhaps local authorities might be persuaded that heat pump installations would provide an advantage. People who are sophisticated enough to understand how to make savings in energy have tended to go fairly well down that road already in things like insulation and as the opportunity for further savings is decreasing, the advantage obtained from a better user of energy is that much less. That it is more expensive than the classic forms of heating installation presents a problem. There has in fact been sales resistance in the north of England to low-energy houses because for such a relatively expensive house the conventional gas heating system was so small.

If it can be demonstrated that heat pumps give an edge in reducing the problems in those houses where there is a condensation problem it is worth pursuing.

D. R. Oughton (Oscar Faber Consulting Engineers)   I would ask the question whether as an industry we are satisfied that the new housing stock has significantly lower rates of infiltration than existing accommodation. I have personal experience of new properties which appear to have good quality window seals but I can report that the rates of infiltration are very significant, which leads to a further question as to whether there is adequate control over standards of construction in buildings of this type.

D. Guy (Denis Guy & Partners)   Twenty years ago I built my house but did not double glaze. Recently, to cut down noise and reduce draughts some double glazing has been installed. The old wooden windows had shrunk and leaked badly at a considerable energy cost. Double glazing should be considered very seriously as an energy saver as well as noise protection.

R. J. Shennan (Max Fordham & Partners)   If a heat pump is chosen as a heating source with boiler supplementation then additional plant space will be required. Further space penalties are incurred with any kind of wet heating system such as radiators, fan convectors, etc. because of the increased size caused by the lower hot water flow temperature (up to two to three times larger). These facts will not be pleasing to many architects and must be considered with sensitivity. The same problem does not occur with an air system. Assuming an air heating system has already been decided, the additional size of an air coil does not have such great space and indeed cost implications. Reversed cycle also becomes a possibility with some summer cooling at minimal additional cost.

An important aspect of using a water source as both heat source and heat sink is that roof-mounted cooling towers and/or air blast condensers can be dispensed with – a positive architectural benefit in many schemes.

2 Building and heat pump: integration by design

D. R. Oughton and J. P. Quick

Introduction

This chapter discusses the interaction between the building, the heat pump and the heating system, demonstrating the need for integrating these basic elements. The brief includes load balancing, heat sources and sinks, heat storage, heat pump hardware and design of the building envelope and orientation to maximize desirable characteristics; for new and retrofit buildings. Many of these subjects are topics for other chapters. This chapter, therefore, attempts to provide a general understanding of the considerations building services engineers should be aware of during concept design.

A limitation of the brief for this chapter is that it should concentrate on heating only applications. In addition, for practical purposes it is necessary to impose other constraints, which are:

- to limit considerations to the British climate.
- to restrict the subject to systems having heating loads in excess of 30 kW, i.e. non-domestic, and generally to commercial office applications with central machinery.

The extent of the problem of full integration will become clear as the various topics are developed; it is not possible, however, to establish optimum design solutions which can be applied generally. Indeed, the constraints imposed upon building services engineers during the design process seldom allow a choice of many key factors.

Integration by design

Perhaps an appropriate definition of the phrase 'Integration by Design' is 'to plan parts together creatively to form a whole'. There are three principal parts which need to be brought together in a heat pump scheme:

- heating load, and how it varies through the season.
- heating plant, its selection and control.
- distribution system, its selection and control.

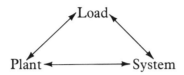

Each of these influence the other two and none can be considered in isolation.

Heating load

Consider the factors and variables which affect the rate of heat loss and the thermal performance of a building; these may be summarized as:

Building shape	– plan dimensions, number of floors, floor height
Orientation	– solar radiation, wind effects
Opaque construction	– area, thermal transmittance, thermal admittance, mass
Windows	– area, single/double/triple glazed, heat reflecting/absorbing glass, internal/mid-pane/external blinds, blind control, infiltration
Internal finishes	– partition type, ceiling type, floor covering

In addition, account must be taken of heat gains arising from internal sources and from solar effects, periods of occupation, external conditions and the level of comfort to be achieved.

To illustrate the variation in heating load which may occur, selected combinations of these factors were used in a computer analysis, based upon a finite difference method of calculation [1], for a building arrangement as Figure 2.1. Figures 2.2 and 2.3 are typical plots selected to illustrate the variation in load imposed upon a heating system during a season. Common to all commercial buildings, the heating load varies significantly throughout the day and from day to day. The system and controls must, in consequence, be capable of responding to this variation by controlling heat input to the space in a regulated and economic

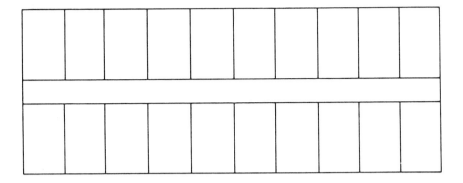

BUILDING MODULES : 3·6 m WIDE X 6m DEEP
                     : 140 TOTAL (70 OF EACH ORIENTATION)
GLAZING, EITHER : SINGLE, CLEAR, 25% OR
                     : DOUBLE, KAPPA FLOAT, 86%
INTERNAL LOADS : LIGHTS 15 W/m²; 08·00 TO 18·30 HR
                     : PEOPLE 1/10 m²   09·00 TO 17·00 HR
                     : MACHINES 3 W/m²
OCCUPATION : 5 DAYS / WEEK
ORIENTATION : NORTH / SOUTH FACING, OR
                     : EAST / WEST FACING

Figure 2.1 *Building details used in analysis of thermal performance and energy prediction.*

manner. This is perhaps an obvious statement, but so often the viability of schemes depend upon this simple factor.

Table 2.1 presents results from the computer simulation showing the variation in heating load with change in building orientation and glazing type. For the limited options tested, a building orientated with north/south facing windows shows a reduction in load over a season when compared to the same building form east/west facing. The results also demonstrate the effect of glazing type on heating load; the glass areas selected for the 'single' and 'double' glazed options have equivalent steady state heat losses, but different solar characteristics.

Table 2.1 also includes results for the same building form glazed to the standard incorporated in the BRE Low Energy Office at Garston. This building is north/south facing with 30% glass area on the north façade and 45% for the south; the windows are double glazed and the areas were determined to give minimum primary energy use consistent with maintaining a comfortable internal environment. Opaque sections

Table 2.1 *Heating load variation with building orientation and glazing type*

Orientation/ glazing	East/west single/clear 25%	North/south single/clear 25%	East/west double/kappa 86%	North/south double/kappa 86%
October	4724	4171	2162	1820
November	6986	6202	4165	2875
December	15,754	14,715	12,536	10,172
January	19,028	17,523	15,076	12,114
February	18,688	17,955	13,376	11,762
March	11,103	10,505	4698	4332
April	3070	3555	96	497
TOTAL (kWh)	79,352	74,627	52,109	43,573
	94,146*			

* System control: As below, but without trv's.

Orientation/ glazing	North/south double/kappa 'BRE'	North/south double/clear 'BRE'
October	1810	2279
November	2473	3606
December	7987	10,347
January	10,265	12,735
February	10,675	13,106
March	5137	6564
April	1013	1468
TOTAL (kWh)	39,360	50,105

System control: optimum start; flow temperature compensated to outside air temperature; terminal thermostatic radiator valves.
'BRE' glazing area: 30% north facing, 45% south facing.

of the façade are of moderately heavy construction with high insulation levels thereby reducing temperature variation and introducing delays of about 6 hours. For heated only buildings, and in the absence of a further detailed analysis, construction to these standards would be a sensible objective.

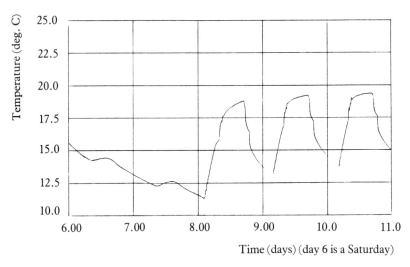

Room dry resultant temperatures – February

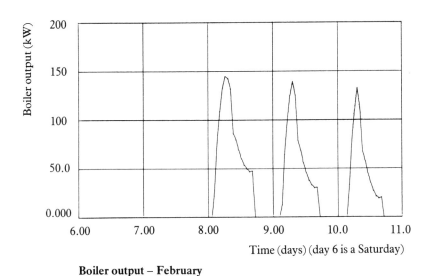

Boiler output – February

Figure 2.2   *Temperature and heating load profiles: single glazed east facing module, February.*

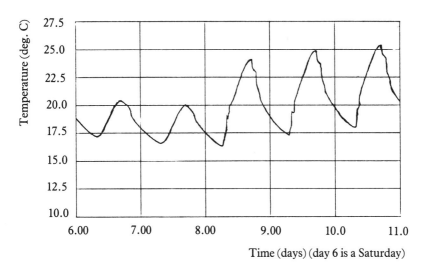

Room dry resultant temperatures – April

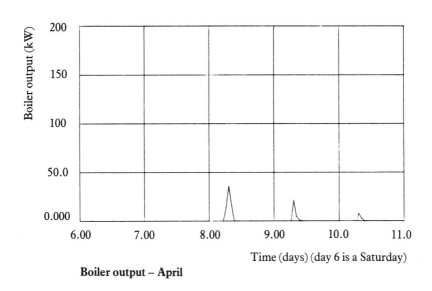

Boiler output – April

Figure 2.3 *Temperature and heating load profiles: double glazed north facing module, April.*

In addition to ensuring that the building envelope and the thermal properties of the building elements are consistent with providing an energy efficient building, the designer must concentrate on producing a system which can effectively match and respond to the variation in heating load.

Heat pump performance

There are four basic heat pump types, which may be defined according to the available heat source and the choice of system:

Source	System/sink
Air	Air
Air	Water
Water	Water
Water	Air

The heat source may be external to the building using a natural source, or internal involving heat recovery from a process or system. Normally, ambient heat sources vary in temperature depending on the time of day or season, whereas internal heat sources may be constant.

The grade of heat available at the source will directly affect the efficiency of the heat pump and in consequence it is important to understand the order of this variation:

- outside air -10–$30°C$
- ground water 8–$12°C$
- surface water 2–$6°C$
- ground coils 6–$12°C$

Any intermediate heat exchange circuit which is required between the sink and the heat pump, perhaps in the case of a water source for protection of the plant against a risk of corrosion, will reduce the grade of heat available at the heat pump, and in consequence reduce the efficiency of the plant. Water source systems which require a brine or glycol/water solution for protection against risks of freezing, will require increased flow rates to achieve heat transfer rates equivalent to a water system, the order of increase is likely to exceed 10%.

The system operating temperature also affects efficiency of operation; the lower the sink temperature the more efficient will be the heat pump. Heating system delivery temperatures may be within these ranges:

- warm air heating   30–50°C
- warm water floor heating   30–50°C
- warm water radiators (low temperature)   45–55°C
- warm water radiators (forced convection)   55–70°C
- hot water radiators (free convection)   60–90°C
- district heating (warm water)   80–100°C

Figure 2.4 was drawn from data provided by a heat pump manufacturer [2] for an air to water device and shows the order of variation in heat output and coefficient of performance. Clearly, both output and effi-

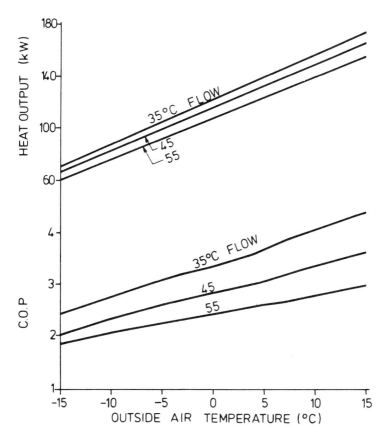

Figure 2.4   *Variation of heat output and COP in air-to-water heat pump.*

ciency ranges need careful consideration during the process of system and plant selection.

Heat pumps which have been developed for heating only applications may have efficiencies greater than those developed as reversible machines for dual heat and cooling purposes. As an example, a reversible machine may have a COP of 2.0 whilst a heating only device may have a COP of 2.4 for equivalent operating conditions; Figure 2.5 gives a comparison over an extended operating range.

It is important to recognize also that the operating characteristics of

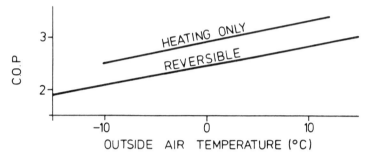

Figure 2.5   *Comparison of performance between reversible and heating-only heat pumps.*

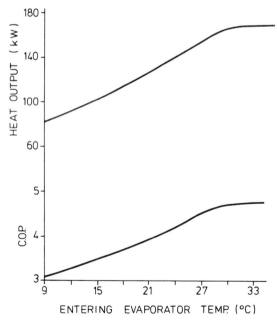

Figure 2.6   *Example of performance plateau for water-to-water heat pump.*

heat pumps are such that the performance, in terms of both output and COP, may reach a plateau on increasing source temperature. In consequence the full range of operating conditions should be examined and not only the characteristics at design temperatures, and it is not recommended to extrapolate the performance curves beyond the manufacturer's published limits. An example of a performance plateau for a water to water machine is given in Figure 2.6.

Plant arrangement

The choice of heating plant must be related to the predicted energy requirements and the cost of fuel and power, the capital cost of plant, design risk in respect of the number of units, that is standby provision, and on the ability of the plant to satisfy the heating load under extreme conditions. Further consideration has to be given to capacity control and the efficiency of the plant under partial load conditions.

There are four plant arrangements which may be considered:

- monovalent – heat pump or boiler
- bivalent (or multivalent) – simultaneous
- bivalent (or multivalent) – alternating
- bivalent (or multivalent) – alternating/simultaneous

The operation of these arrangements is illustrated in Figure 2.7 which shows representations of heating load against outdoor temperature. Although simplistic, these are adequate for the purpose of illustrating the principles of application.

The variation of heat pump output with change in source temperature makes it essential for the capacity of the plant to be established under conditions of maximum load. Air source plant is critical in this respect. The requirement for defrosting air source evaporator coils during cold weather may account for a loss in heat output for up to 10% of the total running period during 'frost' conditions. With an associated reduction in efficiency under cold ambient conditions, air source heat pumps are not suitable for monovalent systems. Water source systems are not normally subject to the same range of low temperatures and in consequence these may be considered for monovalent systems.

It is unlikely in the present economic climate, however, that an economic case can be made for a monovalent heat pump system. Normally, the heat pump would be selected to provide between 50 and 75% of the maximum heating load, the balance being provided by a supplementary heat source. In other than perhaps the domestic situation, consideration should also be given to provision of standby by installing multiple units, or a multi-circuit heat pump.

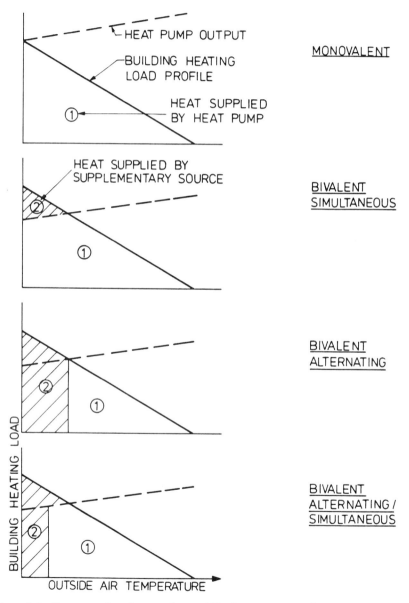

Figure 2.7   *Representation of monovalent and bivalent system operation.*

Potential noise problems arising from the use of heat pumps must be considered early in the design process. Care is required where air source units are proposed and particularly where these operate during night hours.

Manufacturers' literature must be examined carefully when selecting plant, since there are a number of limiting features that require careful consideration.

- When not in operation, limit the temperature flowing across the condenser coils to prevent migration of the refrigerant to the evaporator which may cause damage to the compressor on start-up.
- There is often a minimum sink return temperature limit on a system, which may lead to the need for a supplementary heat source during the initial heat-up period.
- Ensure that the sink flow and return temperatures do not exceed the stated maximum, normally of the order of 10 K.
- Ensure flow rates are maintained above the minimum for safe plant operation.

In respect of bivalent systems, it is important to distinguish between the relative capacities of the two heat sources and the respective predicted period of operation. For this purpose it is useful to refer to an 'S-curve' graph relating heating demand, energy requirement and days per heating season [3].

A definition of a bivalent system, offered in a recent technical article [4], reads thus:

'The effective and efficient operation of two fuels simultaneously, with inbuilt flexibility to change between one and the other so that the maximum use of their respective supply availability and cost advantages can be achieved.'

The need for flexibility in use is often overlooked at the design stage. A case can often be made for the installation of a bivalent fuel source in new and refurbishment projects, but not necessarily incorporating heat pumps. With a view to the future, however, when the case for heat pumps could be more attractive, it would be prudent to arrange for the plant and systems to be compatible with a change to ambient energy-based systems.

When considering the features of flexibility in a bivalent system a combination of a storage fuel (oil or coal) with a direct supply fuel (electricity or gas) is perhaps the best option.

The amount of heating energy supplied by the heat pump and by the supplementary source in a bivalent system will depend upon the system type (simultaneous or alternating) and the selected balance or change-over point. Typical figures have been published for domestic applications [5] based on continuous operation, and a heat pump COP of 2.4, which may be taken to be representative of other applications.

System balance temperature (°C)	Percentage of energy supplied by supplementary source	
	Simultaneous	Alternating
7	14	–
6	10	–
5	7.8	55
4	5.4	42
3	3.2	29
2	2.1	20
1	0.8	13
0	negligible	8
−1	negligible	5

Clearly, these values will be affected by hours of occupation, levels of internal gain and solar benefit. A similar analysis for heated commercial buildings is being developed by the author using computer simulation methods.

It is important to provide effective control to the heating source(s) and associated systems. Plant controls may be considered under two headings: unit protection and plant operation.

Heat pump protection will normally include high and low refrigerant pressure cut-outs, compressor delay start timer, high temperature cut-out, system flow switches and defrost in air source units.

Operational controls will include a means of plant capacity control to match space loads efficiently and to avoid excessive cycling, time-switch or optimum start controller, and bivalent sequence control, with particular care given to limiting use of supplementary heating to times only when required. Inefficient use of supplementary heating is perhaps the most common fault in heat pump controls giving rise to running costs in excess of predictions.

Systems

The choice in system selection is between air and water systems. Air can be discharged into the space through relatively small outlets but requires large distribution ducts and there is a risk of noise from fans or terminals. Water systems require only small pipes but relatively large emitter surfaces.

In smaller systems, a single-zone air system is perhaps an appropriate choice, but the relatively low supply temperatures for economic plant operation, around 35°C, require large ducts which are in consequence difficult to route through buildings. For commercial-scale buildings, water systems allow flexibility for multi-circuit control, which in turn leads to the opportunity of operating the circuits at different temperatures.

The multi-circuit and multi-temperature concept allows the possibility of using more than one circuit to serve a space. One example is the use of a fast-response fan and coil system (operating at 'high' temperature) to supplement an embedded floor coil system (operating at 'low' temperature). This arrangement can take full advantage of low temperature distribution whilst overcoming the problem of slow response inherent with embedded systems. The 'low' temperature system may be used to provide a base load, perhaps 60% of the heating requirements, supplemented by the fan and coil system to provide effective temperature control.

In low water capacity systems, it may be necessary to incorporate a buffer tank to optimize plant operation providing better temperature control and reduced plant 'cycling'. Use of a buffer tank may be necessary for these reasons, quite apart from use for heat storage purposes when demand is out of phase with supply, when low energy cost is out of phase with demand, or for standby purposes. The advantages of heat storage must be balanced against increased capital cost and spacial requirements.

Integrating a heat pump into an existing heating system presents a number of problems. Conventional systems are designed to operate at about 80°C flow and 70/60°C return, with heating surfaces selected to satisfy the peak winter conditions. For much of the heating season the emitting surfaces are not fully 'loaded' and may therefore be supplied with lower temperature water during these periods. To assist in establishing the temperature at which the system should be operated to satisfy loads at times other than at the design conditions, use may be

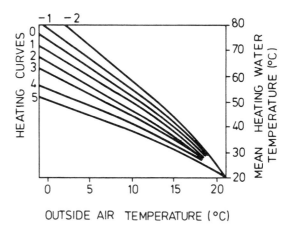

Figure 2.8   *Heat emissions related to water temperatures.*

made of a series of plots given in Figure 2.8. In this simplistic representation, line '0' is the design condition relative to outside air at −1°C. If during tests on an existing system the particular operating condition corresponded to, say, line '1', then it could be shown that for a system operating with 55°C flow and 45°C return, the heating demands of the building could be met by the existing terminals down to an outside air temperature of about 10°C.

Energy analysis

To determine cost effectiveness of a heat pump system, some form of predictive technique is required which can estimate the typical annual running costs of various systems. Predictive techniques are also useful to check the suitability of selected plant – for example: What fraction of the annual heating demand will be met by the heat pump? What size of heat pump is most suitable?

   One technique is the 'heating analysis chart' or 'S'-curve [6] shown in Figure 2.9. This is a graphical representation of the outside air temperature/frequency relationship. Each chart applies to a specific location. The vertical axis has two scales; one is the 24-hour mean outside air temperature, the other is rate of heat loss (assumed to be a simple linear function of outside air temperature). The horizontal axis is scaled in days per annum. The plotted curve shows the cumulative frequency distribution of days at or below a particular outside air temperature. The area under the curve is therefore a measure of degree days. The

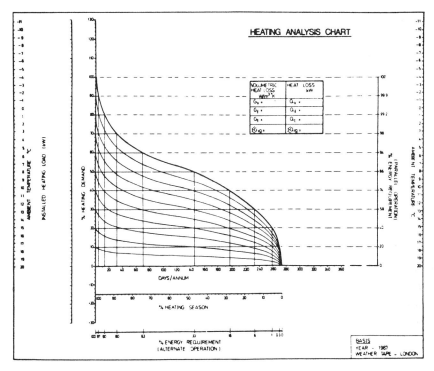

Figure 2.9   *The heating analysis chart: a graphical representation of the outside air temperature/frequency relationship.*

area under the curve may also be used as a measure of total heating energy consumption per year.

The heating analysis chart is used in conjunction with the calculated design heat loss for the building (Figure 2.10). The chart enables the designer to estimate the useful contribution from the casual gains and solar gains. The remaining area under the curve is a measure of the total heating plant energy demand per year. The relative contributions from bivalent heating plant can be established provided that the maximum heat output from each unit is constant or is a simple function of outside air temperature. Parallel or alternating operation may be studied with these charts.

Heating analysis charts are useful as a predictive tool at the early design stage of new buildings. In continuously occupied and heated buildings they can be used with reasonable confidence at most stages of the design. However, if the charts are to be used for the analysis of intermittently heated buildings, correction factors will be needed. This

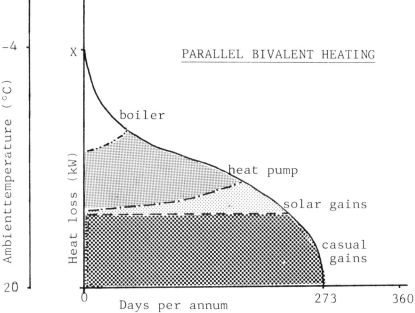

X   is the design heat loss for an inside/outside
temperature difference of 24°C.

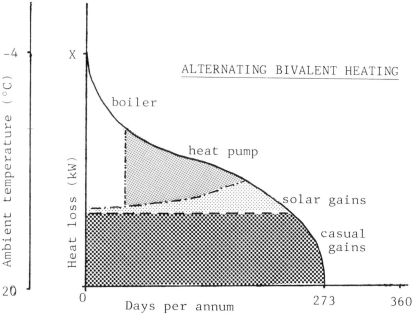

Figure 2.10   *Calculated heat loads for a building.*

is because the heating demand during the preheat period is usually more dependent on the thermal storage effect than on the mean outside air temperature. A single chart can only be used to analyse the mean behaviour over the whole day. To allow for different conditions at different times of day it is necessary to use a new chart for each set of conditions.

Because the heating analysis chart has a number of limitations, it may sometimes be necessary to use more detailed predictive techniques. An alternative analysis method is available which is far more detailed than the S-chart and does not share its limitations. This technique is called computer simulation.

Computer models are used by the Meteorological Office for weather forecasting and by the Treasury for economic forecasting, and simulation techniques are a powerful tool for predicting building energy use and as an aid to making design decisions.

The computer simulation models the thermal response of the building and the behaviour of the heating system as these vary with time and outside conditions. The simulation uses hourly measured values of outside temperatures, solar intensity and other climatological data.

Figure 2.11 shows the computer predicted room temperatures over a

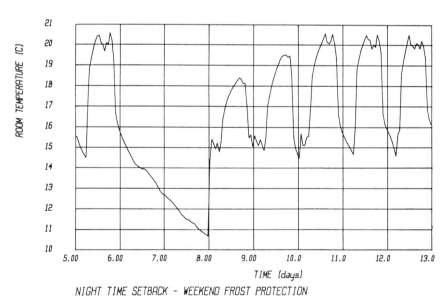

Figure 2.11 *Computer predicted room temperatures over a period of several days: room response in a heavyweight building.*

period of several days, including a weekend. The simulation has high-lighted a problem – the heating has insufficient margin to bring the building up to design temperature after a weekend shutdown. The most energy efficient solution in this case probably would be to reduce the thermal mass of the building, *not* to increase the heating capacity of the system. Other architectural design decisions affecting energy consumption can also be investigated using computer simulations.

Figure 2.12(a) shows some of the many factors which need to be considered. Using a computer simulation the influence of any of these factors can be investigated. Consider item 3; solar gains and casual gains. To predict the beneficial effect of these 'free' gains it is necessary to decide how much of the free gain is beneficial and how much will lead to overheating. This is dependent on the choice of controls. If the controls were very basic responding only to outside air temperature, then overheating would be caused at times of high solar gain. If the controls included a room temperature sensor and some form of feed-back to limit heat emission then an energy saving could be achieved. However, the problem is not as simple as it seems at first sight.

Consider an example: A continuously occupied building with warm-water radiator heating. The construction is to modern standards with wall U-values around 0.6 W/m²K and with 25% single glazing. The radiators are fitted with thermostatic radiator valves.

Table 2.2 shows the results of predicting space heating demand for such a building. Using a heating analysis chart the predicted demand is 416 GJ over the entire heating season. The computer simulation shows very good agreement with this figure when the proportional band on the TRVs is 1°C. However, most TRVs have a wider proportional band – typically 2 to 3°C, but possibly as much as 5°C. Increasing the propor-tional band to 4°C in the simulation produces an increase in heating demand of more than 25%.

Figure 2.12(b) shows some of the many factors which the engineer needs to consider and which can be investigated using computer simu-

Table 2.2   *Predicted heating demand*

	Building 2 (continuous)
Heating analysis chart	416 GJ
Computer simulation	
TRV proportional band 1°C	423 GJ
TRV proportional band 4°C	536 GJ

lation. The importance of controls cannot be over-emphasized – they are probably the single most important influence on energy consumption in a well insulated building. Item 3, plant part-load efficiency, is needed if fuel consumptions are to be calculated.

Of particular interest is the output from an air-source heat pump. Consider two buildings of identical construction; one intermittently

(a)   **The Building Design**

1.   The design heat loss from the building

2.   The thermal response of intermittently heated buildings

3.   Free gains - eg. solar gains and heat from lights

4.   Occupancy pattern eg. hospital : continuous
                                     office   : weekdays

(b)   **The System Design**

1.   System controls eg. thermostatic radiator valves, optimum start

2.   Choice of heating plant eg. heat pumps, boilers

3.   Part load efficiencies of heating plant

4.   Heat output available from the heat pump - particularly significant when outside air is used as the source

Many factors interact with each other

   eg. T.R.V's can react to free gains in the space.

Most factors interact with the climate

   eg. An outside ambient air source heat pump has its minimum heat output when there is a maximum building space heating requirement.

Figure 2.12   *Factors influencing energy consumption.*

EXAMPLE

	Building 1	Building 2
Occupancy :	office hours weekdays only	24 hours every day
Heating hours :	9.00 - 17.00 plus optimum start	continuous

Buildings 1 and 2

System design :   Warm water radiators with T.R.V's.

Oil fired modular boiler plus air-to-water heat pump (rated output 66% of design load)

Figure 2.13   *Example buildings*

heated and the other continuously heated (Figure 2.13). The heating plant is bivalent, consisting of a modular oil-fired boiler and an electrically driven heat pump using outside air as the heat source. Alternating operation is assumed.

The results, given in Table 2.3, indicate that the space heating demand for an intermittently heated building is far higher than would be expected simply by considering the number of hours that the plant is operating. Also, the results indicate a very high degree of sensitivity to the selection of the switch-over point between the two heating devices. When the heat pump is allowed to work down to 2°C outside air temperature, it can provide more than half the required heating energy.

Table 2.3   *Computer simulation results*

	Building 1 (intermittent)	Building 2 (continuous)
Total heating demand (GJ)	303.9	423.4
Percentage supplied by heat pump:		
Bivalent switchover at 2°C	51%	61%
Bivalent switchover at 4°C	34%	37%
Bivalent switchover at 6°C	16%	17%

Table 2.4   *Computer simulation results: system fuel consumptions*

Predicted fuel consumption (in GJ)

Bivalent switchover temperature	Building 1 (intermittent)			Building 2 (continuous)		
	Heat pump	Boiler	Total	Heat pump	Boiler	Total
2°C	57.1	177.3	234.4	95.1	198.6	293.7
4°C	36.9	240.2	277.1	56.0	322.6	378.6
6°C	17.0	307.1	324.1	24.6	432.1	456.7

Simply raising the switch-over by 4 K reduces the useful output of the heat pump to around 16%. Note also the difference between the intermittently occupied building and the continuously occupied one.

The system fuel consumptions (Table 2.4) are particularly useful because they can be converted into comparative running costs and used to estimate the payback period or net present value for a heat pump system.

Two methods of energy analysis have been described, but there are others including one based on monitored performance of existing heat pump schemes referred to elsewhere in the paper.

The two energy analysis methods described should, however, be seen as complementary to each other since each has a place in the design process.

An example of integration

The foregoing text has outlined many aspects of building and services design which require consideration when integrating heat pump systems into buildings. A unique example of heat pump technology combined with an available heat source and with heating system and plant design is the Nuffield Heat Pump. The original system was installed at the College around 1959. After operating for at least ten winter seasons, the heat pump was effectively abandoned as a working unit in the early 1970s [7].

The original plant comprised a heat pump, having an output capacity of approximately 127 kW, arranged to operate in parallel with conventional oil-fired boiler plant as a source of supply to the hot water space heating systems within the college buildings.

The dual compressors and water circulating pumps were driven by a commercial diesel engine and ancillary equipment was provided for recovery of waste heat. Plant configuration overall, however, required that one large circulating pump be electrically driven and the associated consumption of 7.75 kW from the public supply imposed a relatively high penalty upon the plant performance in terms of both primary energy ratio (PER) and operating cost.

The unique features of the original plant were, first, the use of raw effluent from a main public sewer as the base heat source and, second, the configuration adopted for connection of the water circulation systems within the college to the bivalent heat supply.

The fuel consumed by the diesel engine was to the same specification and unit tariff as that used to fire the boiler plant and thus a direct comparison of cost-in-use was practicable, operation of the heat pump showing a relatively significant saving in oil consumption when operational. When oil prices escalated in the 1970s however, the original boilers were removed and replaced by units burning North Sea Gas with the result that the economic comparison was distorted.

When the prospect of rehabilitation first arose in 1973, the concept of all-electric drive from the public supply was considered. Despite the concomitant advantages of much reduced maintenance demands, the penalties in terms of primary energy ratio were such that this approach was abandoned. A parallel solution using a single gas engine, with heat recovery, driving an alternator to provide a power source to all components was investigated. Although attractive in many technical respects, this was not pursued in view of the scale of the installation overall and the requirement that the work undertaken should be confined to rehabilitation, rather than to development of an alternative concept.

In 1978, Ford developed a derivative of their production line 1.6 litre spark-ignition engine modified to run on North Sea Gas. The preferred speed was of the order of 45 rev/s matching a shaft power output of approximately 27 kW: a compact waste heat recovery unit was in the process of development for mounting to the cylinder block in replacement of the conventional exhaust manifold.

The new heat pump machine in consequence consists of two such engines, works mounted as a packaged unit complete with control equipment. One engine powers a Carlyle compressor at 30 rev/s, absorbing 21.5 kW, and the other powers a 25 kVA alternator absorbing approximately 14 kW on normal running load. The alternator is sized to take account of the starting current required by the large sewage circulating pump, powered previously from the public supply. Figure 2.14 illustrates the principal characteristics of the plant arrangement.

The sewage quantity circulated is such that the heat extracted at the rated duty does not result in a drop in sewage temperature of more than 5 K, a requirement imposed by the abstraction licence granted by the Oxford City Council.

The heating systems within Nuffield College buildings require an energy supply of approximately 400 kW when the external temperature is 0°C. The output of the original heat pump was only about one-third of this total. The systems were designed, at the time of the construction of the college, to ensure that the annual demand imposed upon the heat pump machine remained at a constant level for the longest practicable period. They were thus arranged in three equal and separate circuits which, for design external conditions, are:

- embedded panels requiring flow water at 49°C.
- 'low temperature' radiators requiring flow water at 60°C.
- 'normal temperature' radiators requiring flow water at 71°C.

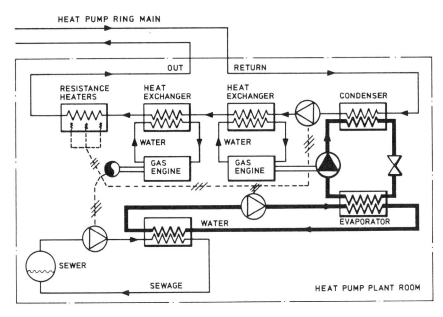

Figure 2.14 *The principal characteristics of the new heat pump machine used by Ford.*

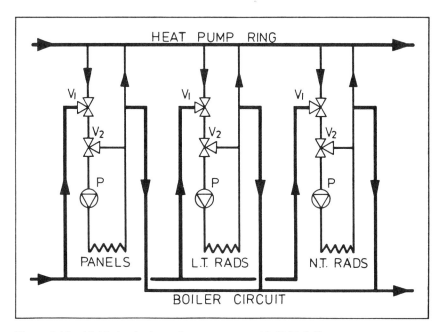

Figure 2.15 *Multi-circuit pipework arrangement at Nuffield College.*

The three circuits are connected to the bivalent heat source as shown in Figure 2.15 and it will be noted that each is provided with two motorized mixing valves connected in series. The first valve in each case is arranged to utilize water flow from the heat pump supply in preference to that from boiler supply, provided that the former is at a suitable temperature. The second valve in each case operates in a more conventional manner, mixing source water with that returned from the heat emitting surfaces. Thermostats are fitted in each circuit to see that water at too high a temperature cannot return to the heat pump ring.

The purpose of this arrangement is to provide a system in which the demand imposed upon the heat pump remains relatively constant, at design level, over the long 40-week heating season required by the British climate for residential occupation. The sequential manner in which the circuits operate produces the load curve illustrated in Figure 2.16 which shows also frequency data taken from meteorological records.

The original diesel engine drive machine, when tested in 1961, had a PER of 1.68 based upon the calorific value of the fuel oil used. This was degraded to 1.25 when the electrical drive from public supply to the large circulating pump was taken into account. For the rehabilitated machine, the heat pump proper would have a calculated PER of 1.82 but drive to the ancillary components would, as predicted, reduce this to 1.42 overall, a value nevertheless almost twice that of the associated boiler plant.

The performance of the new machine will, as a result of co-operation with British Gas, be monitored by the author in conjunction with the Department of Engineering Science at the University of Oxford. British Gas has installed a micro-computer on site, complete with a visual display screen and a cassette unit data store.

The monitoring facilities available are such that the performance of the new heat pump machine can be fully analysed. This analysis will, initially, be applied to the three-system connected load as described previously. It must be emphasized however that a unique facility exists for on-going research into the performance of a heat pump connected either to an embedded panel system, a low temperature radiator system, a medium temperature radiator system, or to any combination of these. Properly exploited and with the co-operation of Nuffield College, an immense amount of data related to heat pump load characteristics and bivalent system operation could be gathered.

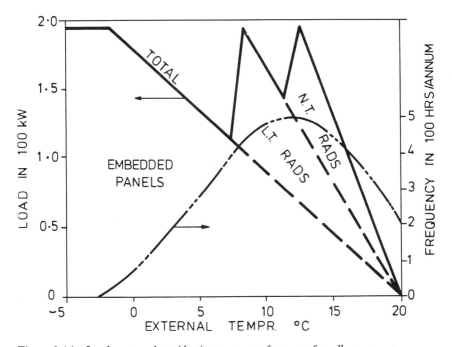

Figure 2.16   *Load curve and outside air temperature frequency for college systems.*

References

1   Facet Ltd. APACHE thermal analysis and energy prediction computer program.
2   Carlyle air to water heat pumps (30GQ).
3   Braham G. D. (1979) Towards Easier Heating Systems Analysis. *Building Services*, March 1979.
4   Morris D. (1984) The Philosophy of Bivalent Systems for New and Existing Buildings. *HAC*, May and June 1984.
5   *Heat Pumps for Domestic Applications, HVCA Guide to Good Practice*, March 1983.
6   See 3 above.
7   Martin P. L. and Kell J. R. *The Nuffield College Heat Pump*. JIHVE January, 1963.

Discussion

Dr A. F. C. Sherratt (Thames Polytechnic)   John Quick in presentation showing the graph of room temperatures over a weekend with the heating switched off (Figure 2.11) suggested that it would be more economic to build a lighter weight of building than to increase mechanical services. This indeed may be true for that period of the year, but life is complex and there may be other periods of the year where having a heavier building might avoid heating altogether, whereas a light weight building would have a heat requirement. A more complex analysis may be necessary to make a total assessment of the optimum fabric. To make proper use of computer simulation it is necessary to know the level of sophistication of the analysis being used and the basic assumptions. Very complex analysis is possible using computers, but it is also possible to present computer-generated results produced from programs little more sophisticated than hand methods and often this is not made clear.

John Quick (Facet Limited)   In general, I agree with most of the comments made by Dr Sherratt, although I do have certain reservations.

In the case of weekend cooling, it is generally true to say that intermittently heated buildings should be thermally light-weight because the running costs will be lower than with heavy-weight structures. However, it is a complex problem. For example, in the particular example, the room temperature did not drop below 10°C even during the weekend shutdown, because of the heat stored in the building fabric. Consequently, a frost protection thermostat, if one were fitted, would probably not switch on additional heating during the weekend. However, in a light-weight building, frost protection heating would be necessary throughout most of the weekend.

So it can be seen that the choice of light-weight or heavy-weight structure has different implications for energy consumption at different times and for different circumstances. One cannot simply say that light-weight buildings are more economical to run than heavy-weight (or vice versa). This is precisely why computer simulation is so worthwhile. Using computer simulation it is possible to look at the behaviour of the building and its systems over the whole year and to evaluate the energy implications of various design decisions. If necessary, a simulation can be run for a complete year, hour by hour.

There are obviously limitations with computer simulation, as with any technique. The greatest problem is occupant behaviour, because people's behaviour is so variable. For example, one person may open a window when they feel too hot, whereas another would turn off the radiator – such factors seem impossible to build into computer simulations. For this reason I consider that computer simulations are more applicable to large commercial buildings than to domestic-scale buildings. In commercial buildings the hours of occupancy are largely predetermined and 'average' behaviour is more likely simply because there is a large number of people.

I agree that it is particularly important that the end user of a computer prediction is fully aware of its limitations. Despite the limitations, it is a valuable technique.

D. R. Oughton (Oscar Faber Consulting Engineers)   The computer simulation used for the study described in this chapter is a sophisticated program and in consequence quite expensive to run; it is therefore unrealistic to suggest that all building designs be analysed by this method. Perhaps the aim should be to use this level of analysis to more fully understand the thermal characteristics of the building envelope and the response of the services systems. This could lead to improved accuracy for manual calculations which could be developed in conjunction with the 'S' curve analysis or similar methods.

Computer simulation certainly has its place in the design process and should be made available to the practising engineer who may not have such facilities available at his own offices.

G. D. Braham (The Electricity Council)   The 'S' curve heating analysis chart was originally developed from an idea by Peter Kalisher by T. Perera and myself and we used it as a method of predicting the energy consumption in swimming pool dehumidifying systems.* Its biggest success was as a means of communicating ideas and it is one of the best means of communicating results from computer studies or from empirical work. The whole performance of the building and its system can be summarized on a simple diagram. The Electricity Council Research Centre at Capenhurst has recently applied the concept to a steady-state program based on the CIBSC method which has been used recently to analyse the BRE building used in this chapter.** The prediction was within 10% – a good agreement for a steady-state program or perhaps any program bearing in mind that an average year is being compared with a real year. I see in the future the main role for the heating analysis charge as a communication technique, an output from other computer programs, virtually anything that can be simulated or measured can be depicted on the heating analysis chart.

R. Gluckman (W. S. Atkins & Partners)   Several heat pumps installed during the last decade have failed to perform adequately because they were wrongly selected in terms of size and part-load performance. It is essential that sufficient effort is made to analyse the heating characteristic of a building to ensure good matching between the heat source (the heat pump) and the heat user (the building). The use of computer programs for this purpose can in many ways be beneficial.

* Braham, G. D. & Perera, G. D. T. (1980) Energy cost saving in indoor swimming pools. University of Aberdeen, 20–24 July.
** Electricity Council Research Centre, Capenhurst – personal communication.

3 Heat sources selection

A. R. Coleman

Summary

The chapter evaluates both technically and economically a variety of heat sources available for a building heat pump. It shows that sources of a higher temperature are favoured for two reasons. First, because of reduced power consumption due to the lower temperature lift and, second, because a smaller compressor can be used for a given duty.

The chapter looks at both 'natural' sources such as rivers, ambient air, etc. and also 'man made' sources such as building extracts, effluent, etc. Each source is evaluated for temperature, thermal properties of the fluids, availability and practicalities of use.

The chapter concludes that man-made heat sources are preferred where available with particular advantages where simultaneous heating and cooling is required (e.g. pub cellars and beer coolers). Of the natural sources, ground water is preferred technically since its temperature is constant, but availability at any site is not certain without some investment in exploratory work. Ambient air is universally available, but in all other respects is a very poor source for a heat pump.

Introduction

The purpose of a heat pump is to take heat from a source, to upgrade it and to deliver it at a useful temperature.

Some sites have a choice of sources, others may have only one – ambient air. For the ecomomics of the heat pump to be as favourable as possible, it is important to consider the whole of the site and the installation together. The selection of a heat source follows naturally from an understanding of the needs of the site and the opportunities presented. No two sites have the same potential or the same requirements.

The 'natural' sources available to the designer are:

- ambient air
- ground
- surface water – rivers, lakes, seawater
- ground water
- direct solar radiation

Man's activities and requirements for environmental comfort cause waste heat to be generated and lost to the surroundings. The heat is carried in flows of air or water which can be the source for a heat pump. The heat pump is then operating as a heat recovery device. A selection of the 'man made' sources available follows:

- building extract air
- effluent – industrial, town
- dehumidification of air
- heat rejected from cooling systems, e.g. computer suites
- special site specific heat sources

Each of these sources will be examined for its suitability with regard to:

- temperature
- thermal properties of the fluid
- availability
- special requirements
- environmental considerations
- system cost

The effect of temperature on the power requirements and the system capacity

Power requirement

The heat pump has less work to do lifting heat from a moderate temperature than from a low one. The COP_H is an indicator of the work requirement of the heat pump cycle (and an indicator of the efficiency of the machine). It is defined as:

$$COP_H = \frac{\text{Heating output of the system}}{\text{Power input}}$$

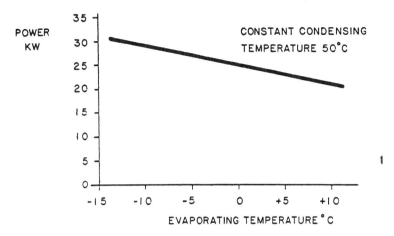

Figure 3.1 *Power consumed to deliver 100 kW of heating from a range of evaporating temperatures.*

Figure 3.1 is a graph of the power required to deliver 100 kW of heating using typical commercially available heat pumps condensing at 50°C, but evaporating over a range of temperatures. It can be seen that raising the evaporating temperature from 0°C to 5°C reduces the power consumption by 10%.

System capacity

The capacity of any heat pump or refrigeration compressor decreases with decreasing evaporating temperature. This is because the compressor has a fixed swept volume but the capacity is a function of the mass flow. A lower evaporating temperature gives a lower density of the vapour at the compressor suction. The mass flow is thus reduced and hence also the capacity.

Figure 3.2 illustrates the substantial decrease in compressor size required for modest increments in evaporating temperature. An increase in evaporating temperature from 0°C to +5°C would typically allow a 17% reduction in compressor size to meet a given duty.

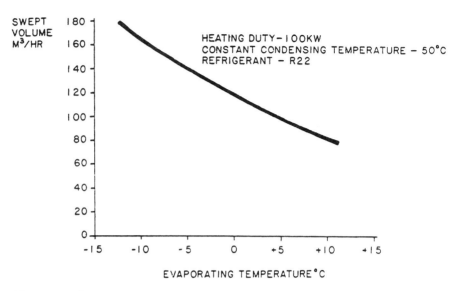

Figure 3.2   *Swept volume of compressor required for a range of evaporating temperatures.*

Some observations on the temperatures of sources

Natural sources, with one exception, fall in temperature during winter. Typical temperature ranges of certain sources are given in Table 3.1 along with corresponding typical evaporating temperatures.

Table 3.1   *Typical winter temperatures of selected heat pump sources.*

Source	Typical winter temperature range (°C)	Typical evaporating temperature range (°C)
Ambient air	−3°C to +10°C	−16°C to 0°C
Ground coils	−5°C	−12°C
River water (and lake water)	+4°C	not below −1°C
Ground water	+8°C to +10°C	−1°C to +3°C
Sea water		
(West coast)	down to +7.5°C	down to −1°C
(East coast)	down to +4°C	down to −1°C
Building extracts	+16°C to +24°C	5°C to +13°C
Effluent	+20°C to +35°C	+5°C to +20°C

Ambient air

Ambient air temperatures in the UK can drop to −20°C in extreme conditions. The dry bulb temperature in London drops below −1°C for some 4% of the heating season and below −3°C for some 1.5% of the heating season. On cold days such as these when the maximum heating duty is needed, the heat pump capacity is at its lowest.

Ground

Ground coils are not affected by short-term shifts in air temperature, however the temperature of the ground does fall progressively over the winter. The ground is a poor conductor of heat and the region around the coils becomes very cold. Moisture migration causes ice formation on the coil which creates a further heat transfer barrier. The temperature of the circulating glycol may fall to around −5°C or below [1], and it will not respond to sudden warm spells.

Surface water

The temperature of rivers and lakes also varies, but less so than that of the shallow ground, dropping to around +5°C [2]. This temperature is the limit to which it is safe to cool water in packaged units without risk of ice-up and tube rupture. A river at this temperature can only be used as a heat source by a special design of heat exchanger.

Ground water

Ground water does not suffer these problems of loss of temperature; the overlying ground insulates it from the seasonal and daily variations in air temperature. The temperature of water in boreholes more than about 8 m deep is roughly constant throughout the year at about 10°C in southern England. In certain areas of geothermal activity, for example localized parts of the Midlands and around Bath, the temperature of borehole water can be significantly higher. Figure 3.3 shows the annual temperature variation in two boreholes and some spot temperatures in the River Thames near Maidenhead [3]. The water from a shallower

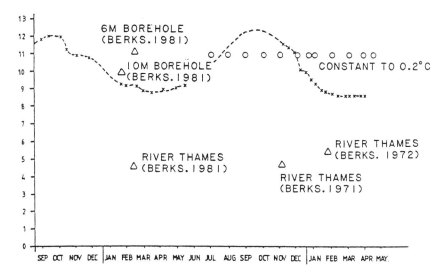

×   5M BOREHOLE IN CAMBRIDGESHIRE (SEP. 1980–MAY 1982)
     (SOURCE–IGS).

○   16M BOREHOLE IN BERKSHIRE (JULY 1971–MAY 1972)
     (SOURCE–IGS).

△   VARIOUS SPOT TEMPERATURES
     (SOURCES–IGS & SMITH & WEBB (DRILLING) LTD.)

Figure 3.3   *Water temperatures in boreholes and the River Thames.*

borehole can show temperature fluctuations but never as much as the ambient air, the ground, or surface waters.

If a heat pump is being considered using a river as the source, it would be advisable to investigate the feasibility of drawing the water from a shallow borehole near the river. The temperature of the borehole water would be higher and not subject to such fluctuation. The water would also be filtered.

Sea

The sea temperature off the east coast of Britain can be expected to fall to 5°C and below. However, seawater is being utilized at Iona off the west coast of Scotland for heating of the Abbey. The temperature here has been reported [4] not to drop below 7.5°C due to the Gulf Stream.

Direct solar radiation

Solar assistance of heat pumps cannot be relied upon on cold days when the capacity is needed. Solar collectors make a negligible contribution in mist or during rainfall; they can only assist another source.

Man-made

People require their environmental temperature to be maintained within close limits compared to the variation in ambient air temperature. The temperature of heat sources originating in that environment, and particularly ventilation extracts, also varies correspondingly little. 'Man-made' sources are thus generally less prone to temperature fluctuations than 'natural' ones.

The effect of the heat capacity of the source

Sensible heat

The heat source must be cooled in order to extract the heat. A source with a low heat capacity flow rate (mass flow × specific heat) must be cooled further than one with a higher heat capacity flow rate. Since the evaporating temperature of the heat pump is always below the leaving temperature of the source, it follows that a lower temperature source with adequate flow can be as efficiently utilized as a higher temperature source with only limited flow available. Figure 3.4 illustrates the temperature profiles through an evaporator for two different sources of different heat capacity.

Wherever seawater is available quantities are limited only by the economic limit of pump size and power consumption. Ground water can be limited by the yield of the aquifer. It is important to design the borehole properly for the prevailing ground conditions and to perform all the 'completing' operations of cleaning and opening the interstices of the surrounding aquifer material to allow water to flow into the borehole with as little resistance as possible.

The heat capacity of the source is determined not only by the quantity of fluid available but also by its specific heat. All common heat pump sources are in the medium of either water or air. Of these, water

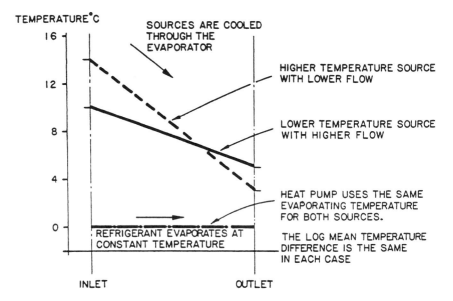

Figure 3.4   *Illustration of temperature profiles in identical evaporators with two different sources.*

has a specific heat four times that of air, the mass flow of water required for a given duty is thus one-quarter that for air. This difference affects the power consumed in transporting the heat source. The evaporator fans of an ambient air source heat pump may consume three times as much power as the water pump of a water source heat pump of a given duty.

Latent heat

In certain specialized cases the source contains latent heat which can be recovered. If saturated air is cooled, water is condensed and gives up substantial quantities of latent heat, leading to higher evaporating temperatures in the heat pump than for a similar flow of dry air. Nearly 2½ times more heat is given up when cooling moist air (25°C, 80% RH) through 10 K than when cooling dry air through the same range. Figure 3.5 illustrates the effect. This enhancement can sometimes be utilized when recovering heat from kitchen extracts and boiler flues. Full use of the process is made in swimming pool heat pump installations.

The water of a heated indoor pool is continually evaporating and the

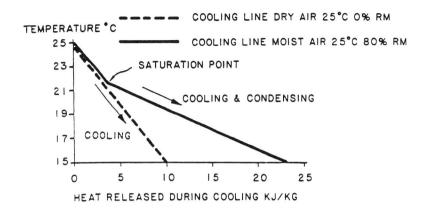

Figure 3.5 *Comparison of heat released during cooling of dry air and moist air.*

humid air has to be ventilated away to prevent condensation and damage to the pool hall structure. The temperature of the pool hall air can be up to 28°C. A dehumidifying heat pump can be used to recover the latent heat of evaporation of the water (as well as some sensible heat) to heat the pool water and the incoming fresh air. The incoming fresh air requirement is also lowered as some of the dehumidified air can be recycled.

The effect of heat transfer coefficient

Heat is extracted from the source material by the evaporation of the liquid refrigerant. Materials with good heat transfer properties allow a higher evaporating temperature to be used or alternatively a smaller heat exchanger.

The heat transfer coefficients of water are an order of magnitude better than those for air.

Availability

The only heat pump source universally available is ambient air. However, unfortunately, it has the worst attributes of any common source in terms of temperature, specific heat, heat transfer coefficient and frost formation.

Ground coils and ground water are the next most usually available.

Ground coils laid horizontally require a considerable ground area to collect heat (30 W/m² ground area). Direct expansion vertical ground coils 10 m deep offer a more compact solution (500 W/m² ground area) [5]. The ice formed around the vertical ground coils may need artificial regeneration during the summer, for example by circulation of water warmed by a solar panel. If the coils reach below a mobile water table this problem will not occur, as the regeneration will be done by the ground water. In such cases however it may be cheaper to operate a ground water heat pump by drawing the water from a well and passing it through a conventional heat exchanger. Ground water from shallow wells is quite widely available. Fifty per cent of the land area of England might be suitable for some site of installation. For a ground-water source device there is often a requirement for two boreholes: for abstraction and recharge. The recharge point would need to be some 2.5 m/kW [6] downstream of the abstraction point to avoid recirculation of cold water (depending on local hydrogeology). Particular care must be taken with water disposal into chalk deposits. In these deposits water flows in fissures rather than by percolation. Recycling is correspondingly more difficult to avoid.

The 'man made' sources are available only at special sites. All mechanically ventilated buildings provide opportunities for heat recovery from extracted air. Heat pumps represent a higher capital cost option than direct heat exchange for these duties, but recover more heat and provide greater long-term savings. Some buildings have requirements for cold areas but are otherwise heated. Ice rinks fall into this category and likewise cold stores and display cabinets in supermarkets. The recovery of condenser heat is in these cases essential for efficient operation.

Another opportunity for heat pumping exists in air conditioning systems with humidity control. The process requires the air to be cooled to its dewpoint for dehumidification and then reheated to prevent overcooling of the space. The reheat duty can be performed by the condensers of the cooling plant.

Ease of use

The heat pump source should be readily usable. It should not require special marterials nor special construction of pumps, etc. to transport it. The fluid should be neither fouling, eroding nor corroding.

A famous heat pump installation at Nuffield College, Oxford, uses town effluent as a source. An intermediate clean water circuit is used to carry the heat from the effluent heat exchanger to the heat pump. It was not thought advisable to use direct expansion (see Chapter 2).

The use of an intermediate circuit adds extra heat exchange and thus extra temperature drop in the system. It is thus a source of inefficiency and should be avoided if possible.

Suitability of sources for capacity

Sources such as the ground, ground water, solar radiation, town effluent, etc. require a greater amount of ancillary equipment and installation works than the simpler sources such as ambient air and building extracts. Smaller installations (less than say 5 kW) are liable to use the simpler sources except in special circumstances.

An illustration of one of the extras for a ground-water installation is the cost of the extraction license (about £130). This is an insignificant cost on a large project but may deter the domestic user.

Maintenance

All heat pumps require maintenance by skilled personnel and experience from the refrigeration industry suggests a cost of 5% of the capital cost per annum. The water well of the ground-water heat pump system requires additional maintenance by different specialists; all that is normally required is an annual clean out and pump overhaul. The cost is likely to be around £700 p.a. for maintenance of the ground works to supply a 200 kW capacity heat pump.

The relation between source and sink

The temperature lift in a single-stage heat pump is limited by considerations of compressor pressure ratio and absolute delivery and suction pressures. Air source heat pumps cannot economically deliver heat to temperatures much more than 50°C. Ground-water heat pumps can reach 60°C and building-extract heat pumps can reach 65°C.

Special requirements

While most sources are immediately identifiable, ground water suffers the disadvantage that exploratory work, including perhaps test drillings, must be carried out to establish the existence and capability of the aquifer.

Ambient air suffers the disadvantage as a source that when its temperature falls below +5°C the evaporator starts to frost up. Frost is a barrier to heat transfer and results in lower refrigerant evaporating temperatures and reduced capacity. The fall off in capacity occurs precisely when the maximum heating duty is needed. The evaporator must regularly be defrosted – a process which draws power and interrupts the supply of heat to the building.

Environmental considerations

The use of ground coils requires an extensive grid of buried piping and can lead to persistence of frost in the soil.

The external fans of air source heat pumps can cause a noise nuisance, particularly in rural villages with low background levels.

The water from a groundwater heat pump should be recharged into the ground to prevent depletion of the resource.

System costs

Unit costs

Costs for heat pump units vary widely depending on the manufacturer and on the type of unit, for example air/air, water/air, etc. Capital costs per unit heating duty for package heat pump units are in the approximate range £250/kW for domestic units (<10 kW) to £50/kW for larger (≈250 kW) water-source units. There are few package ambient air source units of the 250 kW size but the indicated price is however around £100/kW.

Installed costs

Costs for complete installations vary even more widely than the costs for the units only. The circumstances of the site are of critical importance.

Some figures are given below for a selection of recently installed heat pumps. The prices are intended as a guide only.

System 1 – ambient air to air – darts hall
Reverse cycle rooftop units [7],
ducted supply air system included in price.

Nominal heating capacity (15°C air temperature)	91 kW
Total cost	£28,600
Cost/kW	£314/kW

System 2 – ambient air to water – outdoor pool
Heat pump heating water in outdoor
swimming pool, summer operation only.

Nominal heating capacity (15°C air temperature)	92 kW
Total cost	£19,500
Cost/capacity	£212/kW

System 3 – borehole water to water – museum
Heat pump supplying 800 m² floor area
of underfloor heating.

Heating capacity	50 kW
Total cost (incl. well)	£11,100
Cost/capacity	£222/kW

System 4 – air to air – building extract heat recovery

Nominal heating capacity	200 kW
Total cost	£24,400
Cost/kW	£122/kW

It is stressed that the cost of heat pump installations depends on the circumstances of the site.

In general heat pumps using man-made sources are cheapest since the source temperatures are the highest. Very low or nil capital and running costs can be argued for heat pumps that are primarily performing a necessary cooling duty and where use is made of condenser heat. Systems that fall into this category are:

- computer suite cooling
- beer and cellar cooling in pubs
- ice-making machines in hotels and supermarkets
- reversible air-conditioning systems, used for heating in winter

Table 3.2   *Attributes of sources*

Sources	Temperature	Thermal properties	Special requirements	Environmental considerations	Cost	TOTAL
Natural sources						
Ambient air	−	−	−	−	○	−4
Ground coils	−	−	○	−	−	−4
Surface water	○	+	−	○	+	+1
Ground water	+	+	−	○	○	+1
Solar assistance	○	○	−	−	−	−3★
Man-made sources						
Building extracts	+	○	○	○	+	+2
Effluent	+	+	−	○	○	+1
Dehumidification	+	+	○	○	+	+3
Heating/cooling	+	+	+	+	+	+5

★ A heat pump cannot be fully solar source.
+, Good; ○, Average; −, Poor.

Conclusions

The attributes of heat sources are summarized in Table 3.2.

The selection of a source for a heat pump requires careful consideration of all the circumstances of the building. There may be advantageous sources such as ground water or town effluent which are not immediately obvious.

In general, man-made sources are preferable having higher and more constant temperatures. The ideal situation is where a simultaneous heating and cooling load exists in the building and the balance between the supply and extraction of heat is close.

Where a natural heat source has to be relied upon, the first question should be 'Is ground water available?'. No other natural heat source retains its temperature in winter as ground water does.

References

1   Private communication from the Institute of Geological Sciences.
2   Edmunds W. M., Owen M. & Tate T. K. (1976) *Institute of Geological Sciences*, Report No. 76/5, HMSO 1976.
3   See 1 above.

4   Smith G. T., Ian G. Lindsay & Partners (1983) 'Heat Pump Installation, The Abbey, Iona' from *Energy Saving in Building* – proceedings of the International Seminar held in The Hague, November 1983.
5   Goulbourn J. R. & Fearon J. EEC Contract EEA-4-035-GB.
6   Coleman A. R., W. S. Atkins & Partners (1982) *Ground Water Source Heat Pumps for Space Heating in the UK*. Report prepared for the Department of Energy under Contract E/5C/CON/2684, 1982.
7   Private communication from HCE Services Limited, Southampton.

Discussion

G. W. Aylott (The Electricity Council)   Mr Coleman has stressed the benefits of ground water supplies, but are there constraints on their use either because of legal requirements or additional costs imposed by the water authorities?

A. R. Coleman (W. S. Atkins & Partners)   The water authorities are supportive of any application to use ground water and are particularly interested in this system. Some water authorities believe that there is a requirement to obtain an abstraction licence for domestic systems. It will certainly be necessary to obtain an abstraction licence for a commercial system. The disposal of the water is then subject to the consent of the authority but no cost is incurred. An abstraction licence application costs approximately £130, and it is a one-off cost. The cost of the water then abstracted – provided it is recharged back into the aquifer – will be only about £30 per year for a building of approximately 400 kW heating capacity. Licence fees are therefore a minimal part of the cost. There may be a problem in some areas where the water is not readily available but provided it is proposed to recharge the water back into the water table and not deplete the water resource, then the water authority is likely to look favourably on an application.

R. J. Shennan (Max Fordham & Partners)   If it is proposed to use a pair of wells as both heat source and heat sink, is there any future in considering the reversal of the roles of suction and recharge wells during summer and winter, in order to obtain some benefit from the effect of local heating up or cooling down of the source during the cooling and heating seasons respectively?

A. R. Coleman (W. S. Atkins & Partners)   Borehole water at 10°C is a much better condensing medium for the rejection of heat than air at 25 or 30°C. Using borehole water to satisfy a condenser cooling load will therefore result in lower running costs for any air conditioning system. There is a possibility that heat rejected into the aquifer can then be re-utilized in the winter. Such a system would require careful engineering to ensure that the heated water is recoverable. There has been much discussion about using aquifers for thermal storage purposes but commercially it has yet to be proved feasible.

J. Nisbet (J. Nisbet & Partners)   With the increasing use of electrical energy in buildings for computers and other high technology equipment, is it possible in the future, by using heat pumps, to make a building self-sufficient and independent of an external heat supply?

A. R. Coleman (W. S. Atkins & Partners)   Any mechanically ventilated building presents opportunities for heat recovery from the extract. Typically a large proportion of the energy requirement for a highly insulated building is for heating the ventilation air. The casual gains from lights, occupancy and activities can provide all the heating that is required without an external heat source. Heat pumps are normally required in such buildings, though in some cases direct heat recovery is adequate. Cold start up of self-heating buildings should be given attention at the design stage.

4 Selection criteria for heat pumps

D. P. Gregory, J. W. Kew and G. Hamilton

Summary

This chapter discusses the difficulties of selecting a heat pump for domestic or small commercial applications, and the problem faced by the designer or user who is presented with different, non-comparable information by different heat pump manufacturers. The steps already taken by the German and French authorities in establishing standards for describing the performance of heat pumps are reviewed.

The selection of a heat pump

Heat pumps provide one option for the heating, and/or cooling of buildings, and possibly the provision of hot water. The heat pump option must be examined in the light of alternative heating or cooling systems, which are termed here 'conventional' systems. There are two broad questions that must be answered.

1　Why use a heat pump rather than a conventional system?
2　If the heat pump is the right answer, how does one specify the right one for the job?

This second question may further be subdivided as follows.

1　What type of heat pump is most suitable for the application?
2　How should the correct size or capacity of the heat pump be selected?
3　How will the manufacturers' stated performance data relate to the performance obtained in a real application?
4　How can the energy consumption or running cost of a proposed heat pump installation be predicted?

In this chapter guidance is given towards obtaining the answers to these questions, and gaps in current information are identified which make some of the answers difficult or impossible to obtain.

Access to information

Over the past ten years BSRIA has been involved in assembling published information on heat pumps, producing bibliographies, conducting market studies and applications studies, preparing recommendations for a European Heat Pump Standard, and answering enquiries from member companies. A huge growth in the amount of published material concerning heat pumps has been observed. For example, Table 4.1 shows the number of abstracts contained in three successive BSRIA bibliographies and Table 4.2 shows the cumulative total of heat pump abstracts contained in the IBSEDEX database compiled by BSRIA.

Table 4.1 *BSRIA bibliographies* [1, 2, 3]

Date issued	Number of abstracts
1975	270
1981	505
1984	500*

* Highly selective

Table 4.2 *IBSEDEX database*

Date	Cumulative total of heat pump abstracts
Jan. 1983	750
July 1984	2120

A recent technical note [4] on heat pumps listed no less than 45 different UK manufacturers and distributors, many of whom made available their technical literature for the purpose of preparing this chapter.

Contained in this wealth of literature, most British published experience relates to air to air reversible systems which our market studies [5] indicate account for over 58% of the UK market. There is surprisingly little published British experience on ground or water source heat pumps of any type, and of air to water heat pumps designed for heating-only applications.

Water source heat pumps are uncommon in the UK, although large areas, especially in the south and east, do have underground water sources at about 8°C all year round [6].

Over 46% of abstracts on the IBSEDEX database are in foreign languages. Overseas, different types of heat pumps predominate in the market place, but information on cost and cost effectiveness is difficult to interpret because of fluctuating exchange and inflation rates, different fuel costs, different tax incentives and different climatic conditions.

Why choose a heat pump in the first place?

A heat pump is usually considered because it can offer a reduction in energy consumption, usually synonymous with a reduction in operating cost. There are other special features which encourage the selection of a heat pump: the same equipment can provide heating in a building designed to be air conditioned, or the special features of heat pumps to recover waste heat at higher temperatures can be exploited. In general, five factors have been identified which have a favourable influence on the cost-effectiveness of heat pumps. The nearer an application complies with these factors, the greater the chance of an economically and technically viable design.

1 An opportunity to use cooling or dehumidification and the production of useful heat simultaneously.
2 A constant rate of heat requirement and a constant temperature differential between source and supply.
3 The ability to satisfy the heat requirement at a low temperature.
4 A need for air conditioning or cooling, and space heating, at separate times but in the same building.
5 The need to conserve fuel due to high cost or the unavailability of lower cost, conventional alternative fuels.

Before selecting a heat pump merely on the grounds of energy conservation, consideration must be given to the cost-effectiveness of other means of reducing energy consumption. Freund [7] has listed the internal rate of return for various methods of conservation. He does not include heat pumps in his list because 'they do not show a saving in recurrent costs compared with gas heating in the UK'.

Similarly, before selecting a heat pump on grounds of reduced running cost, comparison must be made with the opportunity to take advantage of off-peak electricity tariffs, at the cost of providing storage capacity.

Selection of heat pump type

There are many different types of heat pump, classified by fuel supply, heat source and heat sink carriers, and whether or not they are reversible. In most cases the choice of heat pump type is forced externally by fuel availability, or design concepts within the building to be serviced. As a generalization, for space heating applications, and at present fuel prices, an electric heat pump is only cost-effective in comparison to oil-fired heating [8]. If gas is available, a gas-fired boiler will usually be the less expensive alternative. This situation could change if the technology of gas engine-driven or gas-fired absorption cycle heat pumps develops.

If the facility of air conditioning (available from a heat pump at little extra capital cost) is exploited, one must be cautious of specifying air conditioning in a building where comfort cooling is not essential. This results in an overall increase in energy consumption, simply because the designer chooses the use of the cooling capability 'because it is there'.

The building owner or operator may however choose to provide comfort cooling, at little extra capital cost, and consider the additional running costs of summer cooling to be worthwhile in terms of customer or staff comfort.

As far as the heat distribution system is concerned, there appears to be considerable public resistance to warm air heating (it has a poor image as being noisy, dirty and dry), and large spaces are requried for ductwork, especially for the low temperatures associated with heat pumps. On the other hand, radiator heating is popular in the UK and modern double panel, finned convector units are quite suitable for the lower temperatures involved. In 1983 BSRIA [9] estimated that of 12 million domestic central heating systems operating in the UK, 75% were wet (radiator) systems. In a poorly insulated building it is unlikely that one would select a heating-only heat pump without upgrading the building fabric insulation first or at the same time.

Selection of the correct size of heat pump

The sizing of a heat pump is a far more complex and critical process than sizing a conventional boiler system. For conventional boilers in domestic applications there are fairly standard procedures [10, 11, 12], which are: to estimate the heat loss of the building at an outdoor

temperature of $-1°C$; to add about 10% to cover intermittent operation; to add 2.0–3.0 kW for domestic hot water; and to install the next
larger commercially available size (installers will often add a small
'safety' factor to this to ensure that the boiler will really be big enough).

Table 4.3 shows the capital cost of heat pumps in comparison to
boilers and resistance heaters [13]. In general, the capital cost of a heat
pump is about three to four times that of a boiler, so that over-
estimation of capacity cannot be afforded.

Table 4.3 *Capital cost of heat
generators*

Heat generator	Capital cost £/kW
Gas boiler	55
Oil boiler	85
Air-to-air heat pump	210
Electric resistance heater	35

Oversizing will also cause excessive cycling of the heat pump, which
may reduce its performance and will certainly shorten its life.

Unlike a boiler, the heat pump itself should not be sized to balance
the heat losses for an outside design temperature of $-1°C$. The heating
capacity of an air source heat pump reduces with increased 'temperature lift', or with decreasing outdoor temperature, so to avoid huge
overcapacity at average winter conditions, low temperature conditions
are dealt with by providing supplementary or auxiliary heating. In
principle, the sizing exercise is as follows. Figure 4.1 shows the heat loss
from the building as a function of outdoor temperature. Overlaid on
this is a performance curve of capacity versus outdoor temperature for a
heat pump, drawn here as a straight line for simplicity. The crossover
point is the 'balance point' below which auxiliary heating must be
provided. The designer can select different sizes of heat pumps which
will result in different balance-point temperatures. The recommended
balance-point temperature is a matter of divergent opinion. It is interesting that the HVCA *Guide to Good Practice* [14] for domestic heat
pumps does not make a precise recommendation, but instead offers
tables showing a range from $-1°C$ to $7°C$, and states that 'The installing
engineer should advise the customer of the ambient temperature at

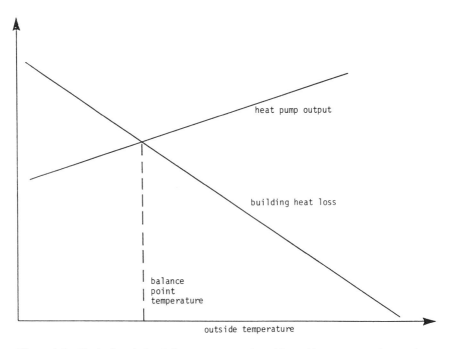

Figure 4.1   *Typical variation in heat pump capacity with outside temperature is opposite to that of building heat loss.*

which the heat pump can just meet the requirements of continuous operation (the balance point)'. This emphasizes the need for standardized test methods to provide the designer with reliable performance information over the heat pump's operating range.

The guide adds: 'Because of the poor defrosting performance at lower than normal temperatures, subject to manufacturers' recommendations a heat pump should not be run for prolonged periods in cold weather at low flow temperatures'.

It is appropriate to mention here that the CIBSE has published a technical memorandum (TM 11 *Selection and Application of Heat Pumps*, 1986).

The capacity of the supplementary heating must be capable not only of maintaining the heat loss at the 'outdoor design temperature' (normally −1°C in the UK), but must provide the capability of rapid recovery from cold in the event of a system shutdown. Supplementary heating may be provided by a conventional boiler if one already exists in a retrofit situation, but for a new installation electric resistance heat is cheapest and normally used. This leads to another major area of concern for the system designer: if the heat pump is switched off at night, the morning recovery will take place at the expense of resistance heat at a COP of 1.0, whereas if the heat pump operates all night it can supply all or part of the necessary heat at a COP of maybe 2.0 or more and probably at off-peak electricity prices. Thus it may actually cost less and use less energy to leave the heat pump on than to switch it off, and the HVCA Guide [15] suggests that 'consideration should be given to the operation of a heat pump on a continuous basis'. Night temperature set-back rather than night shut down is probably the best answer to this control problem.

In the case of a reversible air to air system, the cooling capacity is usually matched to the building's requirements. In commercial buildings this will normally result in a heating capacity 30–40% below requirements [16]; supplementary resistance heating would be used to make up the difference.

The Electricity Council has played a leading role in developing techniques for the sizing and control of reversible air to air heat pump systems and supplementary heaters. Its extensive monitoring exercises, particularly for high street shops, have determined the effect that these factors have on the overall running costs [17]. They offer several recommendations.

1   There is a need to ensure that the output of the heat pump meets the building heat loss at a temperature of 2–3°C.
2   Adequate preheat periods are essential to ensure the heat pump alone provides most of the energy that the building requires and to minimize the need for supplementary heating with a subsequent lowering of the maximum demand incurred on cold days.
3   If fresh air is not required during preheating periods, the fresh air damper should close automatically.
4   When supplementary heat is required it should be staged in discrete quantities as the outside temperature falls to minimize the effect of maximum demand charges.

Applicability of manufacturers' information for the system designer

In order to 'size' the heat pump, capacity data must be available for at least two standard sets of conditions, so that the way in which capacity varies with outdoor temperature would be known. More importantly, for the designer to be able to choose between one heat pump and another, these data should be presented by all manufacturers under comparable conditions and in a similar form.

The variation of capacity with outdoor air humidity is due to the tendency of frost formation and the effectiveness of defrosting techniques used. Different heat pumps will suffer from frost formation to differing extents, subject to the design of the outdoor heat exchanger, but in general in the temperature range from −5°C to +5°C frost formation will occur and some form of defrost will be needed [18]. Too small an evaporator can cause defrosting to be necessary at external temperatures of up to 13°C [19]. This greatly affects the seasonal COP for a major part of the range of external temperatures under which a domestic heat pump would operate.

Defrost can be provided in several ways: by reversing the refrigerant cycle; by a hot gas bypass in the refrigerant loop; or by operating an auxiliary electric heater under the evaporator coil. Each of these methods has very different implications in the overall system design, and the energy required for defrost must be taken into account when estimating the overall power consumption. One cannot expect to be provided with an almost infinite range of data over various temperature and humidity conditions, but for basic design and comparison purposes the power consumption or COP should be provided for at least one condition under which the defrost mechanism of the heat pump is operating, so that maximum power requirements are known.

The power consumption of a heat pump defrost is not only that required by the compressor, but includes that of the heaters and the fans. For proper design, the gross power consumption must be specified, but if auxiliary heaters are built into the heat pump, their power consumption must be stated separately, so that the system designer may better be able to select the balance point at which these heaters become operative.

Prediction of seasonal operating cost

It is difficult enough to predict the fuel cost or energy consumption of a conventional central heating system, and far more of a problem to do this for heating-only heat pumps. A number of computer simulation models have been developed for predicting seasonal heat pump performance and estimation of fuel cost [20, 21, 22] but there are many variables, not all of which can be taken into consideration even in complex models. In practice, the performance of a heat pump is dependent on other factors that cannot easily be modelled by a computer, for example quality of installation, commissioning and maintenance.

This emphasizes the importance of site monitoring of heating-only heat pumps where the decision to use a heat pump would be based solely on potential fuel cost savings.

Electricity Council monitoring of reversible air to air heat pumps, already discussed, provides guidelines for likely seasonal operating costs particularly for the air conditioning of high street shops, this being the largest application of heat pumps in the UK [23, 24].

There are some important factors which must be known by the system designer if he is to be able to predict running costs. Among these are:

1   The power consumption and anticipated frequency of operation of defrosting systems. This can be as much as 2% of the seasonal

Table 4.4   *Energy consumption of supplementary heaters*

System balance-point temperature (°C)	Percentage of design heat load met by heat pump (A)	Percentage of seasonal heat demand met by supplementary heater (bivalent supplementary)	
		Morgan & McMullan [26, p. 110]	HVCA Guide to Good Practice [14, Table 3, p. 18]
−1	100	0	negligible
0	–	–	negligible
1	90	1	0.8
2	–	–	2.1
3	80	3	3.2
4	–	–	5.4
5	70	6	7.8
6	–	–	10
7	60	10	14
9	50	15	–
11	40	22	–
13	30	30	–

Note (A) – Assuming room temperature = 19°C
　　　　 – Outside design temperature = −1°C

energy use [25]. This figure does not include the energy used by the supplementary heaters which may be operated during defrost to maintain the heat supply.

2　The ratio of seasonal energy supplied as resistance or supplementary heating rather than as compressor energy to the heat pump. This is a most important factor and is related to the selection of balance-point temperature. Table 4.4 shows two similar estimates of this quantity.

Comparison of manufacturers' literature

From the foregoing we can see that in the very least the designer needs to know:

- Output over the whole operating range of external temperatures (for air source heat pumps)
- Frequency and duration of defrosting and the energy required to achieve defrost
- Information from different manufacturers presented in a form which makes comparisons easy

We obtained manufacturers' literature from about 30 different UK distributors and reviewed 20 of these relating to air to air and air to water systems for space heating and cooling applications. Of the 20, only 12 referred to some form of national standards, as shown in Table 4.5.

Table 4.5　*Use of national standards in heat pump data sheets*

Standard	Subject	Number of manufacturers quoting the standard
BS 2582:1970	Air conditioning	5 (cooling mode only)
AR I 240	Unitary heat pumps	2 (heating mode only)
AR I 320-81	Water source heat pumps	1
AR I (unspecified)	–	1
DIN 8900	Heat pumps	1 (heating mode only)
ISO R.859	Room air conditioners	1
BACAB	Heat pumps	1
None quoted	–	8

As a consequence of this lack of uniformity in presenting data, the 'standard' conditions under which capacity and COP data are presented vary quite widely, and make comparisons of one manufacturer's heat pump with another very difficult.

Some manufacturers quote the output heating capacity at one, two or up to seven different outdoor temperatures, usually using 21°C as the indoor temperature and either 35, 45, 55 or 60°C for water outlet where applicable. Some manufacturers give graphical presentations of capacity versus outdoor temperature, without stating which, if any, points on the curve are experimentally determined. Outdoor temperatures used for data points range from −15°C to +24°C in the heating mode, and the very critical data at or around 0 to +8°C is quoted at 0, +2, +5, +7, +7.2 or +8°C apparently at the manufacturer's whim.

A most puzzling situation arises when attempting to determine the expected performance under frosting conditions. In most cases it is unclear whether the quoted performance (capacity) data were obtained under steady-state conditions with the defrost controls 'on', and if they were, whether the energy used in defrost was accounted for in the COP calculations.

Since frosting becomes most severe [27] at temperatures between +5 and −5°C, and less important at the colder and dryer sub-zero conditions, one would expect the energy consumption to rise in this temperature range and thus the COP versus outdoor temperature curve to show a 'dip' at this point. None of the manufacturers' data which included graphical presentation showed such a dip, from which we conclude that defrost energy is excluded from their published data.

Heat pump performance standards

It comes as a surprise to many that there are no British Standards for heat pumps. The British Air Conditioning Approvals Board (BACAB) published a draft performance, rating and safety standard in 1981 [28], and this was issued for comment with a BSI foreword with the intention that it would form the basis of a British Standard. The BACAB document draws largely from air conditioning standards in existence in the USA, and from an international safety standard currently under review by the International Electrotechnical Commission [29]. The performance and rating section of this draft document proposes test methods and test conditions for various types of heat pump in the heating and cooling mode, but only under steady-state conditions where frosting and defrosting do not occur. A system designer in the UK is faced with the problem of not being able to cite a BSI standard test condition or procedure by which the performance of different makes of heat pump may be compared, while the manufacturers face the problem of not having an established and agreed test procedure. Hence they are reluctant to invest funds in an elaborate testing laboratory which might be rendered obsolete or irrelevant by the subsequent publication of a standard.

This problem seems to have been overcome both in Germany and in France. There is a family of DIN standards in Germany [30] relating to testing and peformance rating of air to water and water to water heat pumps. These standards address the specification of a reproducible test laboratory, test procedures, temperature and humidity conditions, and specify calculation procedures including the definition of, and an unambiguous calculation of, the coefficient of performance. Testing under conditions where frosting occurs is accomplished by requiring the test to continue over a period of at least four defrost cycles and integrating the overall energy inputs and outputs. While this is still something of an 'artificial' test condition, it is reproducible from one place to another, and enables a comparison of COPs of one heat pump with another to be made under reproducible conditions.

The French standards are similar in many respects to those of Germany, but they embrace a wider variety of heat pump systems, including the use of extract air as the source, and air to air systems [31]. There is a series of 'experimental' French standards which set mandatory pass or fail levels for extreme operating conditions and for some severe start-up and restarting tests, but these standards have not been adopted yet.

Neither the French nor German standards address performance in the cooling mode. Heat pumps in France and Germany are more often than not 'heating-only' devices, in contrast to the UK situation, where more than 60% of heat pump sales are for heating and cooling applications. There are well established testing standards for air conditioners in the UK [32] as well as in France and Germany, and these can fairly easily be adapted to the cooling mode of heat pumps. The UK does of course also require heating-only performance standards and test methods.

The simple existence of a standard does not solve the dilemma of the systems designer unless manufacturers use the standard at least to describe the performance of their product, if not to actually produce test data. Equipment selectors can induce manufacturers to do this by specifying BSI (or even AFNOR or DIN) performance data. There is a serious problem of cost, however.

The market for heat pumps in the UK is quite small, yet the cost of a fully instrumented and controlled test rig capable of performance measurements to, for example the DIN or AFNOR standards, runs into tens of thousands of pounds, and the test time required for a comprehensive set of COP versus temperature data would be extensive. Thus there is a need to keep testing procedures brief and simple if all manufacturers are to be expected to submit their equipment to test. In France and Germany there are government and industry subsidies available in various forms for the building and operation of independent heat pump test facilities, and such facilities have been built in parallel with the development of standards with step-by-step checking that the written procedures are indeed operable.

Finally, one must consider the selection of temperature conditions used for 'standard' performance ratings. In the heating mode, DIN specifies five outdoor test conditions, ranging from $+24$ to $-18°C$ (dry bulb). AFNOR specifies four or five outdoor test conditions depending on the heat pump type, ranging from $+13$ to $-7.5°C$. It is unlikely that more than two conditions, one at about $+7°C$, which is a frost-free condition, and one at $+2°C$, which is a severe frosting condition at which performance would be integrated over a number of defrost cycles, would be agreed and adopted as an international set of mandatory test conditions. However, two reproducible and comparable test points, under which capacity and COP are reported, are far better than the current array of different test or calculation points.

The CEN (European Committee for Standardization) has set up a technical committee to produce a common European Standard for electrically driven heat pumps. Publication is not expected before 1988 at the earliest. BSRIA has carried out a project for the EEC on comparing the current and proposed heat pump standards of the EEC member states, and recommending the text of a European standard. The report of that work has been given to CEN as a contribution towards European standard.

References

1   Loyd S. (1975) *The Heat Pump*, BSRIA Bibliography LB 103/75 BSRIA, Bracknell, 1975.
2   Loyd S. (1981) *The Heat Pump*, BSRIA Bibliography LB 103/81 BSRIA, Bracknell, February 1981.
3   Loyd S. & Hamilton G. (1984) *Heat Recovery from Buildings*, BSRIA Bibliography 104/84, BSRIA, Bracknell, May 1984.
4   Hamilton G. & Kew J. (1983) *Heat Recovery and Heat Pumps in Buildings*, Technical Note TN 4/84, BSRIA, Bracknell, August 1983.
5   Petts C. (1984) (Ed.) *Statistics Bulletin*, Vol. 9, No 1, 1984. BSRIA, Bracknell.
6   Page P. & King B. (1983) Low energy school design: a case study. *Building Research and Practice*, May 1983.
7   Freund P. (1979) *The cost effectiveness of some measures for energy conservation in buildings in the UK*, Proc. 2nd International CIB Symposium on Energy Conservation in the Built Environment, May 1979.
8   See 7 above.
9   King A., Tregoning G. & Webster J. (1984) (Eds.) *Statistics Bulletin*, Vol. 9, No. 2, 1984, BSRIA, Bracknell.
10   Chartered Institution of Building Services, *Guide Book A, 1978*.
11   *Central Heating: A Guide to the design and operation of domestic wet central heating systems*, British Gas Corporation.
12   *National House Building Council Handbook*, Section S9.10.
13   Barnes J. C. (1983) *Heat Pumps and Maximum Demand*. CIBS Technical Conference 1983. Chartered Institution of Building Services, London.
14   *Heat Pumps for Domestic Applications – Guide to Good Practice*. Heating and Ventilating Contractors Association, London, March 1983.
15   See 14 above.
16.   See 13 above.
17   Goodall E. G. A. (1981) Monitoring heat pumps in use. *Building Research and Practice*, Jan/Feb 1981.
18   Heap R. D. (1983) *Heat Pumps*, 2nd Edition. E. & F. N. Spon Ltd., New York.
19   Bartlett R. J. & Chabra J. R. (1983) *Engineering development of a domestic (air to water) heat pump*; paper to 'Prospects for domestic heat pump' conference at University of Warwick, September 1983.
20   Henderson W. D. The simulation of a heat pump heating a highly insulated dwelling. *Building Services Engineering Research & Technology*, Vol. 3, p. 117.
21   Rosell J., Morgan R. & McMullan J. T. (1982) The performance of heat pumps in service – a simulation study, *Energy Research*, Vol. 6. 83–99, 1982.
22   Vahid F. (1983) Cost analysis of domestic heat pump systems. *Building and Environment*, Vol. 18, No. 1/2, 1983.
23   See 13 above.
24   See 17 above.
25   See 18 above.
26   McMullan J. T. & Morgan R. (1981) *Heat Pumps*. Adam Hilger Ltd., Bristol.
27   See 18 above.
28   *Interim Standard for Rating Performance and Safety of Air to Air and Air to Water*

Heat Pumps up to 15 kW Nominal Output. British Air Conditioning Approvals Board, London, 1982.

29 *Safety of Household and Similar Electrical Appliances – Particular Requirements for Air to Air Electric Heat Pumps*; draft, International Electrotechnical Commission, IEC 335-2-61D (Central Office 9), 1982.

30 *Heat Pumps, Definitions; Testing Conditions, Methods and Markings; Testing of Air to Water Heat Pumps*. DIN 8900 parts 1, 2, 3 and 4, and DIN 8947 Deutsches Institut für Normung e.V. Berlin.

31 *Heat Pumps: Test Methods; Terminology and Classification; Rules concerning fitness for use*; AFNOR E38-100, 101, 102, 103, 104, 105, and 110. Association Française de Normalisation, Paris.

32 BS 2852: 1970 Rating and testing room air-conditioners.

Discussion

N. F. Cross (British Telecom)   Coming from a traditional heating and ventilating background, and conservative about unusual approaches, is this the time to seriously investigate and apply heat pumps or would it be sensible to wait 2 or 3 years?

Dr D. P. Gregory (Building Services Research & Information Association)   The answer must depend on the particular application in mind. For example, to heat an outdoor swimming pool, use a heat pump now, whereas the time to apply heat pumps to heating domestic dwellings is not yet here. The criteria given in Table 4.1 is the principal guideline and those criteria are reflected in many of the other chapters. If an application comes pretty close to any of those guidelines, a heat pump should be considered, but there are rather few applications that do fit. For the average run-of-the-mill heating and ventilating application, think twice before choosing a heat pump.

B. Dann (South East Gas)   Maintenance of heat pumps can be a problem. It is extremely difficult to find suitably qualified service people to look after the gas engine-driven unit described in Chapter 5 without having a series of sub-contractors in site. The requirement is for a competent electrician/ refrigeration/ automotive/ heating engineer and, as intimated, such an individual has been difficult to locate.

G. D. Braham (The Electricity Council)   The application of heat pumps and reverse cycle systems, is a firm goer, but there are other areas too which are worthy of consideration. A number of these have already been discussed,* in particular the use of heat recovery in conjunction with heat exchangers for ventilation systems represents a good energy source. Calculations suggest that the use of packaged equipment is likely to be attractive for this application although the decision between individually designed and packaged systems is always difficult. Quality control in both design and site manufacture and site commissioning can be a problem and this is reduced with packaged systems. Reliability and marketability must therefore be separately considered in assessing packaged versus central systems.

J. W. Kew (Building Services Research & Information Association)
Reliability and life-time of heat pumps are questions continually asked of BSRIA staff particularly by building owner-operators considering using heat pumps. At present, life-time data quoted appears to be the same as that for refrigeration equipment – around 15 years depending on type.

The operating conditions for heat pumps are more arduous and operating hours per year are usually longer than for refrigeration equipment both of which could presumably shorten their life span and reduce their reliability.

Dr J. Paul (Sabroe & Co.)   The oldest large electrically-driven heat pump installation being monitored was built in 1971 and was redesigned in 1980–81 from ammonia, which was common in the days when it was built to R22. It is a screw compressor unit and has never been repaired in the last 7 years – some 30,000 hours of operation. Another engine-driven heat pump in a slaughter house works 24 hours a day throughout the whole year, and is switched off only at Christmas for inspection. Although not serviced, it is looked after by a professional operator – a mechanic who really takes care. Reliability has been such that if the gas engine were to fail, it would be cheaper to buy a new one than to have paid the manufacturer's service costs. This engine has operated more than 30,000 hours in under 4 years.

I am very confident that reliability is obtainable if a heat pump is bought with a close eye on quality and efficiency, not which is the cheapest one. Neither does this mean a preference for the most expensive but rather a sound relation between quality and price.

We won a fixed price tender for repair of a reciprocating compressor working on ammonia, 10 years old but whose working life was unknown, although subsequently discovered to be about 100,000

* Sherratt A. F. C. (1984) (Ed.) *Heat Pumps for Buildings* Hutchinson, London.

hours running. The machine had never been serviced so it had done close to 100,000 hours without moving parts, just belt wear. The crank shaft, the piston rods and the bearings were all in good order. Some new parts were put in to justify the price, but really it was not necessary.

R. Gluckman (W. S. Atkins & Partners)   On conventional refrigeration applications the figures Dr Paul gives are very common throughout the UK. I do quite a bit of trouble-shooting on plant that is pre-war (e.g. old twin cylinder vertical reciprocating machines with large flywheels running at 200 r/min). In the extrapolation to heat pumps one has to be careful. First, in small units, and particularly the domestic ones, anything like that kind of life-time cannot be expected. In the larger units, if the pressure and temperature conditions at the discharge of the compressor are comparable to normal, and that would depend on the refrigerant used, there is no reason to expect that heat pump reliability should not be somewhat similar to that of refrigeration plant. If compressors are being stretched beyond their normal capabilities, this will not be the case. There is an interesting example in comparing an actual installation with a 'light' industrial compressor and one with a more rugged type of machine. The light machine is designed to the limit, for refrigeration operation, and in heat pump operation it underwent a number of failures because of thrust-bearing problems. The unit is being over-stressed in heat pumping and it is not being over-stressed in refrigeration. Hence we must be careful not to use the refrigeration rules of thumb and in many cases I think we are going to be lucky if we get a life of much more than 10 years.

I am very strongly of the opinion that we are getting so many problems because often people who have only a vague knowledge of heating and ventilating systems are responsible for installing heat pump systems. There is much too little education and training on refrigeration (and hence heat pumps as the two topics are so similar) in the whole university system.

Dr A. F. C. Sherratt (Thames Polytechnic)   Training is important and perhaps in the past insufficient emphasis was put on heat pumps and refrigeration, indeed refrigeration courses have been quite limited. Today, degree courses in building services or environmental engineering are available which give a good fundamental base in refrigeration as well as other aspects of building services. There are however insufficient places to satisfy the industry and the industry has never been good at recruitment at the higher levels.

5 The gas engine heat pump installation at the Purley Way Service Centre

V. Sharma and B. Dann

Introduction

Since its arrival on the UK energy scene, natural gas has enjoyed evergrowing popularity, in spite of the many changes that have occurred in the market place. This success has derived to some extent from the Gas Industry's ability to adapt and to exploit the many advantages that the new fuel had to offer over its predecessor and also over many of its competitors. In particular, the relative cleanliness of the products of combustion led to greater flexibility in flueing and the inclusion of a wide variety of heat recovery devices without fear of undue corrosion. With continuing interest in fuel economy, two more developments have come to the fore in the past few years: condensing boilers and heat pumps. The former, a relatively new concept, offer the prospect of extracting all but a very small proportion of the heat possessed by the fuel. The latter, by no means a new concept, can deliver more heat than that contained in the fuel. It is this unique prospect of heat multiplication, or something for nothing as perceived by some, that has captured the imagination of certain sections of the public to such an extent that heat pumps are viewed as serious rivals to the wide variety of energy efficient hardware already on the market.

Whilst the response of UK manufacturers has been cautious, their EEC counterparts have been particularly active in driving the new technologies to the market place. In anticipation of successful transition to full commercialization, the SEGAS Regional Management Committee decided in 1981 that it would be prudent for personnel at the operational levels to gain first hand design and installation experience at the earliest opportunity. Two condensing boilers, an absorption heat pump and a gas engine-driven heat pump were installed in three SEGAS premises. Each system was monitored to assess duty performed, fuel efficiency, servicing requirements, reliability and economics in order to ascertain scope for commercial exploitation.

This chapter deals with our experience with the gas engine heat pump providing space heating at the Purley Way Service Centre. The building chosen was the first available on the SEGAS building programme. It was not necessarily the best available on economic considerations. Construction phase had just commenced and it was just possible to redesign the heating system to accommodate the heat pump and the associated hardware. An air to water heat pump was chosen because of the preference in the UK for wet central heating.

For the sake of completeness, an attempt has been made to highlight all the relevant aspects of the learning process, even though some of them now, with the wisdom of hindsight, may appear a matter of common sense. This chapter also describes some of the thought processes which occurrred at the time, albeit necessarily subjectively.

Design considerations

The building to be heated was a single storey service centre of approximately 800 m² floor area (Figures 5.1 and 5.2). The largest proportion

Figure 5.1   *Purley Way Service Centre.*

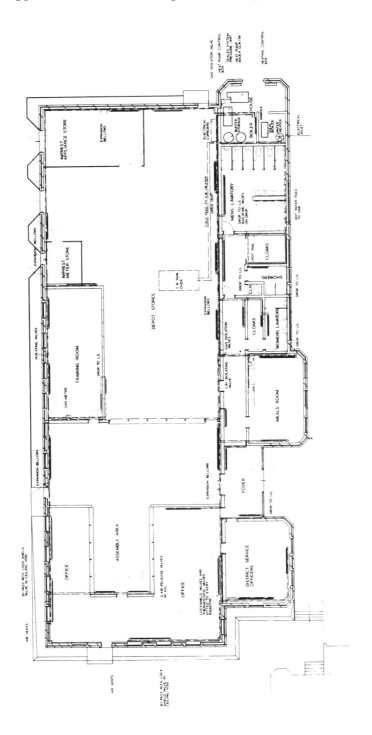

Figure 5.2   *The layout of Purley Way Service Centre.*

of the space was taken up by a fittings store, with three offices and a manual workers' communal area, and usual services such as kitchen and toilets completing the layout. The heating equipment was housed in the plant room at the west end of the building. It can be seen that the space required to accommodate the heating equipment is well in excess of that which would be considered to be the norm for conventional boiler plant. Further, it was necessary to strengthen the roof structure to support the weight of the evaporator unit, and the boiler had to be isolated from the heat pump for safety considerations in the event of a refrigerant leak. These obviously represent additional costs which could often be overlooked at the design stage of a new installation.

The steady-state design load was estimated to be 64 kW for the design conditions of $-1°C$ outside, internal air temperatures of 16.5°C in stores, and 19°C elsewhere. There was an element of 'art' in this estimate because an assumption had to be made about the ventilation rate in the store room which varied considerably depending upon whether or not the loading bays were open. This had a positive aspect, however, that the provision of the plant overcapacity for preheating at start-up was not excessive as the ventilation loss overnight was very much less than that assumed for steady-state daytime operation.

A survey of the UK market for a suitably sized air to water heat pump revealed that the hardware available was still at the prototype stage; that it was aimed at a market in excess of 100 kW; and that it was unlikely to be available in time to meet the contractural obligations dictated by the main contract. A wider search identified production units made by Bauer of Germany based on the Cortina engine and Bock compressor (see Appendix I). The manufacturer made very impressive claims regarding thermal performance, sound levels and field experience, albeit all of it in German conditions. The price of the unit, including transportation costs, was lower than for units likely to be available in this country.

From this point on, the design of the heating system was directed by Bauer's German experience, adapted where possible to suit the local climatological conditions and design and installation practices. An early consequence of this was the provision of service hot water from a separate appliance. The demand was thought by Bauer to be too low to justify operating the heat pump during summer months for this service.

An early topic of debate was the sizing of the heat pump and whether to opt for a mono- or bivalent system. German experience suggested a bivalent system with the heat pump sized to provide 40% of steady-state load. It was suggested that for the UK conditions a figure between 50 and 60% might represent the optimum on economic grounds.

More careful consideration of the factors underlying this balance point revealed that at least four significant factors suggested raising the heat pump proportion for the very much milder UK climatological conditions ($-1°C$ design temperature against $-15°C$ in Germany). These were:

1   The refrigerant characteristics would permit the heat pump to operate at all ambient temperatures encountered in the UK. In Germany heat pumps do not operate near the lower end of the temperature range, forcing the choice towards bivalent systems.

2   The drop in the output of the heat pump near the ambient design temperature was lower over the narrower UK temperature band.

3   The loss in fuel economy due to defrosting as design ambient temperature was reached was unlikely to be significant.

4   A simple analysis of fuel consumption showed that only a small proportion of fuel is burnt at outside air temperatures above that corresponding to 50% steady-state design load. This would severely restrict the opportunity for the heat pump to contribute significant savings to the annual fuel bill.

Further, as the cost differential between a 40 kW and 64 kW was only £3000, that is, less than 8% of the total installation cost, it was decided to proceed with the unit rated nominally at 64 kW (at 5°C). A gas-fired boiler, rated at 96 kW, was installed to supply the excess demand and to serve as back-up in the event of off-line maintenance.

Bauer insisted on the inclusion of a buffer thermal store (2000 litres) to minimize excessive cycling of the heat pump during mild weather to ensure good fuel economy and enhance reliability of the hardware. These clearly need to be balanced against the additional costs of the storage and the standing heat loss from it (Figures 5.3 and 5.4).

The control system ensured that the heat pump operated at one of the two preselected engine speeds chosen, between 1150 and 2300 r/min, to meet building demand for heat. When this had been achieved, the thermal store was charged up until the return limiting thermostat switched off the heat pump. The thermal store continued to supply the heat until its output was insufficient, when the heat pump switched on automatically. The boiler operated only to satisfy the peaks in heat demand when the output of the heat pump was insufficient. This parallel operation offered maximum fuel economy.

Figure 5.3   *Heat pump with evaporator unit.*

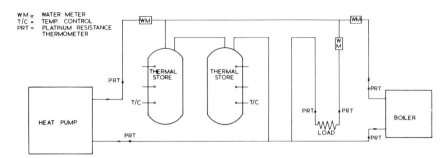

Figure 5.4 *System layout.*

The usual practice of sizing the heat emitters for 80°C flow and 70°C return temperatures was replaced by the manufacturer's recommended 65°C and 45°C respectively. Since the average system temperature dropped from 55 K to 35 K above the design space temperature, the increase in the radiator surface required to maintain the heat emissions was larger than it first appeared. The system was also changed from the usual open-vent to sealed type.

Since the building split conveniently into north and south facing aspects, two compensators and three-port valves were used to control each zone in response to the outside detectors mounted on north and south walls. Thermostatic radiator valves were used as return tempera-ture limiters and not to control the space temperature. The boost facility overrode the normal compensator control to provide rapid heating up in the mornings. Provision was made from the outset to monitor the performance of the heating system in sufficient detail to enable adequate validation of the design assumptions, to gain an under-standing of the working of the system, to fine-tune it to best advantage, and to obtain sufficiently detailed performance data to compare and contrast performance with a conventional boiler system. It was import-ant to steer a sensible course between the two extreme approaches to monitoring of relying entirely on fuel bills, on the grounds that this is the only comparison of interest to the customer, and measuring every-thing that moves as often as possible because that ensures sufficient data for any desired comparison. In practice, the former rarely provides adequate detail to make meaningful comparisons, whereas the cost of gathering, analysing and interpreting the minute detail associated with the latter is invariably out of proportion to the real value of the information.

Keeping the objectives of the monitoring exercise firmly in focus, it was decided to derive energy balances for the four main elements – heat pump, thermal store, boiler and the building load – using high-precision platinum resistance thermometers and water, gas and electric meters. External air temperature and humidity were also monitored, together with internal space temperatures at two locations. In order to speed up data collection and analysis and to minimize manpower costs, a micro-based Commodore 8050 data-logging system was used. All parameters were scanned at 1-minute intervals, averaged on an hourly basis and recorded on a floppy diskette for subsequent detailed analysis and interpretation in the laboratory.

Installation

It was apparent from the outset that the heat pump would impact upon the building structure in a number of significant ways. Mention has already been made of the need to extend the plant room to accommo-date the additional hardware and to reinforce the roof structure to support the evaporator unit. Further, it was necessary to isolate the open flue boilers from the heat pump to ensure that no trace of the refrigerant, in the event of an accidental leak, could enter the boiler combustion chamber, causing excessive corrosion damage. To avoid this eventuality, a separate sealed room was created within the plant room to house the boiler and the service water heater, with the ventila-tion air provided from the outside. The same applied to the engine air intake.

The usual electric supply to the conventional boiler system was supplemented with a three phase supply for starting the engine of the heat pump. This obviated the need for a lead/acid battery and the subsequent maintenance associated with it.

As stated earlier, one of the objectives was for SEGAS fitting staff to gain experience in installing a heat pump. This arrangement provided some useful feedback on a number of aspects. These included the need for more than the usual care in pipe-jointing for the pressurized system, not the norm in this country, and additional manpower required to manhandle the larger radiators, which were virtually double the normal size, to compensate for the lower system water temperatures. In many instances three persons were required to lift and position them on the wall brackets. This process was repeated during the subsequent decorating phase. In-house installation effort had to be supplemented in two specialist areas, namely the refrigerant and electrical aspects. Two local firms were found to possess the desired expertise and were able to accommodate the work to suit the project time-scales. Without any doubt, the majority of the problems encountered related to integrating the separate control systems associated with the heat pump, thermal store, supplementary space heating and service hot water boilers and the usual space heating controls. These arose, in some instances, out of the difference in design criteria and practices in Germany and the UK. After several discussions with the manufacturer, a workable arrangement was produced. Fortunately, the knowledge and enthusiasm of the electrical contractor extended well beyond the routine aspects. This was tested thoroughly on more than one occasion.

These modifications to the original specification by the manufacturer resulted in changing the heat pump operation from single to dual speed operation and using the supplementary boiler in parallel operation, instead of the original alternative of switch-over at a preselected outside temperature. These provided a better match between heat pump output and building demand and considerably increased the hours of heat pump operation, giving better fuel economy.

Performance

The heat pump was commissioned by Bauer at the beginning of April 1982. The weather for the remainder of that heating season was on the whole very mild, making it difficult to run the heat pump for long enough periods to obtain any useful performance data. In the limited operation that was possible, four areas were highlighted for early attention.

The first manifested itself as unexpected switch-over to the supplementary boiler on occasions when neither the heat pump had failed nor the demand for heat had exceeded the maximum output available from the heat pump. The intermittent nature of the fault made it difficult to identify its cause quickly. In the event, it was traced to a lack of engine oil pressure during start-up, causing the pressure-switch and the associated safety interlocks to cut off the heat pump. At the recommended speed of 1150 r/min, the preset period of 6 seconds, during which safety interlocks were bypassed, was not sufficient to build up the pressure necessary to suppress the cut-off switch. Extending the bypass period to the maximum permitted of 10 seconds overcame the problem. The ease of implementing the cure was out of proportion to the effort involved in locating the fault.

The lower speed of 1150 r/min was also thought to be responsible for the second problem, the premature breaking up of the direct drive coupling between the engine and the compressor. At Bauer's request this speed has been abandoned in favour of 1400 rpm. Besides overcoming this problem, the working oil pressure was reached more quickly at this higher speed, reducing risk of lock-out at start-up. To date, neither problem has recurred.

The third problem highlighted in the early days was that of excessively long preheat periods required by the heat pump to raise the building space temperature to the required level. Three factors were responsible for this. First, the heat pump never reached the specified flow water temperature of 65°C. In practice it was usually less than 50°C and rarely exceeded 55°C at the specified flow rate through the heat pump. Second, the mean system temperature was very much lower than that achieved in a conventional wet system, resulting in reduced radiant heat transfer to the building and the occupants. Finally, the thermal capacity of the system was very much larger because of the increased metal and water content of the system. Reasonable preheating times were obtained by utilizing the 'boost' facility of the weather compensator control and allowing the boiler to raise the system temperature, if required, during this period. Incorporation of a boost facility thus becomes essential and not just desirable, as is sometimes the case with conventional systems.

Had the decision not already been taken to provide the service hot water from a separate source for other reasons, the reduced flow water temperature would have been a serious set-back, with a significant cost penalty. In this instance it just provided one more good reason for separating space heating and hot water duties.

Finally, the acoustic performance of the evaporator unit turned out to be something of a disappointment initially, not least because of the low noise levels quoted by the manufacturer due to the unusual fan arrangement. Bauer blamed this on a particularly bad batch of fan motors and supplied replacements from Germany within 3 days. The sound levels reduced to within specified figures and lower than normally expected. To date, there has not been any significant deterioration in performance from the sound standpoint.

One of the features of interest to the design engineer is the variation of the heat delivered by the heat pump with the source temperature, in this case the ambient air. In practice the full output was required only for short periods at start-up because the source temperature also dictated the heating load. These data points – hourly averages – have been extracted and are shown in Figure 5.5. It can be seen that in a real system the variation was quite small. This is due to the practicalities of the heating system and the contributions from the engine water and exhaust gases.

Figure 5.6 shows the thermal performance of the heat pump in terms of the COP for a range of external air temperatures. The definition of the coefficient of performance (COP) is not a classical one but a practical one, preferred by most practising engineers, that is

$$COP = \frac{\text{Heat delivered by the heat pump}}{\text{Gas consumed by the heat pump}}$$

The consumption of electricity is discussed later.

Each point represents daily averages of COPs and external air temperatures, which as mentioned earlier are an accumulation of hourly averages recorded by the datalogging system, which in turn were based on scans of the relevant parameters at 1-minute intervals. Daily aver-

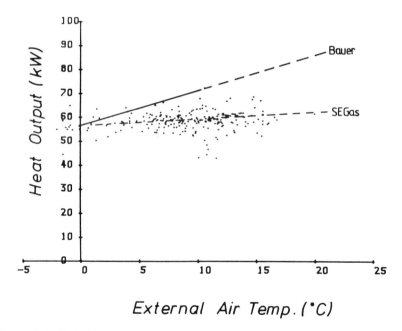

Figure 5.5   *Variation of full heat pump output with external air temperature.*

ages were preferred to hourly or weekly ones because the former were too volatile due to the slow system response and the latter were too crude to give a realistic feel for the behaviour of the system.

Also shown are performance data obtained by the manufacturer. It can be seen that both the heat delivered and the COPs obtained at Purley Way were lower and that variation with the external air temperature was less pronounced. The explanation for this discrepancy would appear to lie in the difference in the test conditions under which the two sets of data were obtained. Manufacturer's data are thought to have been obtained in controlled laboratory conditions. It is well known that under such conditions, performance is often over-estimated because the system does not suffer from the reality of continuously varying climatological conditions which affect the heat demand, and in this case also the output of the heat pump, variable system temperatures, interaction of automatic controls and further complication in this case of the thermal store and the supplementary boiler.

Figure 5.7 shows that the improvement in the heat pump during milder weather, that is, reduced load, compensates for the cycling losses normally associated with heating equipment. These normally

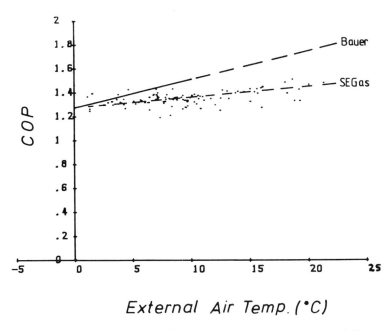

Figure 5.6   *Variation of heat pump COP with external air temperature on daily basis.*

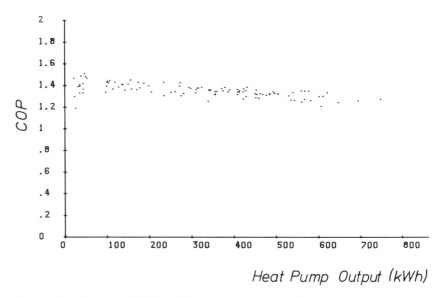

Heat Pump Output (kWh)

Figure 5.7   *Variation of COP with heat pump output on daily basis.*

appear as reduced efficiency at part-load. In the case of the heat pump there is a small increase in the COP, despite these losses.

Nevertheless the fuel utilization performance of the heat pump is impressive, somewhat better than normally expected. This is due to the large contribution from the heat recovered from the engine cooling water and the combustion products.

Seasonal COP has been computed to be 1.33, showing a fuel saving of just under 44% over a conventional gas boiler. These figures drop to 1.25 and 40% respectively when the standing losses from the thermal store and the power consumption of the additional electrical equipment associated with the heat pump are taken into account. Standing loss from the store has been estimated to be 14 kWh each day whilst the additional electricity consumption was 5 kWh or approximately 20 kWh equivalent of gas saved. The standing loss may be less significant in situations where heat requirement is continuous, especially as the lower overnight temperatures will increase the overall fuel consumption by a greater proportion than the increase in hours of operation. The electrical consumption will increase in direct proportion to the number of hours the heat pump is on. Fuel used for defrosting purposes was negligible.

Throughout the heating season there was only one instance of the

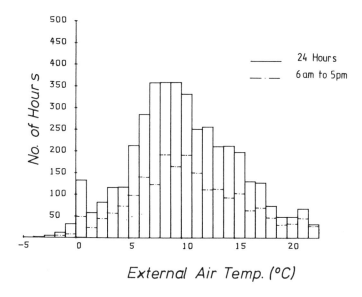

Figure 5.8   *Distribution of external air temperature at Purley Way Service Centre.*

defrost operating and that only for a matter of minutes. This is not particularly surprising in view of the mild winter conditions that prevailed during 1983/84 (Figure 5.8). The few hours of sub-zero temperatures occurred overnight when the heat pump was not required to operate.

Advantages of an engine heat pump are therefore clear in terms of fuel utilization efficiency. Obviously this is only one element of the equation in assessing the viability of a particular heating scheme. Some of the others are now discussed.

Economic considerations

It has already been shown that the Purley Way heat pump, excluding the thermal store, achieved fuel savings of 44% over a conventional heating system in a complete heating season. This reduced to 40% when the standing losses from the thermal store and the additional electricity consumption were taken into account. There may be some scope for optimizing the size of store to minimize these losses. The latter figure represents a fuel saving of £400 for this building.

Table 5.1 shows a breakdown of the costs incurred for the complete

Table 5.1   *Heating system costs*

1 Supply of heat pump	£11,900
2 Supply and installation of refrigerant lines	£ 1,200
3 Additional electrical work	£ 3,500
4 Automatic pressurization unit	£ 1,200
5 Additional building work	£ 3,000
6 Supply and installation of boiler and hot water heater	£11,000
7 Additional installation work	£10,300
TOTAL	£42,100

heating system at Purley Way. It can be seen that the total installation cost was nearly four times the supply cost of the heat pump. Admittedly, some of the costs were unique to this installation. The additional building work is certainly one such item which, if included at the design stage, may not cost as much. Further, the cost of the pressurization unit only became an additional expense because the original system was of open-vent type, the norm in this country. Similarly, there is a small element of 'learning costs' which are an inevitable part of any first scheme. This is particularly true of a proportion of items (3) and (7). With the experience gained from this exercise these could be reduced by 15% in the future. Exclusion of the installation work associated with the monitoring equipment would result in a further saving of 5% in item (7). This figure still remains high, bearing in mind that there is also an element of installation charges in item (6).

Taking account of the potential cost savings mentioned above, a similar future installation would be expected to cost approximately £36,500, that is £25,500 more than a conventional boiler system. Clearly the fuel savings achieved do not give an acceptable return on investment. The situation is exacerbated further when due account is taken of the higher maintenance costs and a lower life expectancy (10 years) of a heat pump compared to a conventional boiler system. The degree of expertise required to maintain the equipment has been found to be greater than anticipated. Already, the cost of maintenance has far exceeded that expected for an equivalent gas-fired boiler system. Additional space requirement – nearly three times the floor area – must also appear as a cost on the balance sheet. Table 5.2 shows typical heating system costs and expected annual fuel savings using a conventional gas boiler as a datum. For example, a condensing boiler system can be

Table 5.2   *Relative system costs and annual fuel savings*

No.	Heating equipment	Additional system cost (%)	Annual fuel saving (%)
1	Conventional gas boiler	Datum	Datum
2	Condensing gas boiler	60	12
3	Gas absorption heat pump	183	15–32
4	Gas engine heat pump	232	40–44

expected to reduce the annual fuel bill by 12% for an increase in the system cost of 60% in comparison with a conventional gas boiler system. The corresponding figures for a gas engine heat pump are 40% and 232%.

To put the above discussion into its proper context, it should be remembered that this site was chosen because of early availability to gain design and installation experience and not because it was thought to offer the best return on investment. Any conclusions drawn should not be generalized without due appraisal of the relevant factors. Better return on investment would be expected in the following situations:

1   Where demand for heat is larger because the fixed costs would become a smaller proportion.
2   Where hours of heat pump operation are longer as in residential heating or some industrial processes.
3   Where temperature rise required from the heat pump is lower, as in heat recovery situations where the source temperature is high, or where there is demand for low grade heat, as in swimming pools.

There are many installations in these categories in the commercial and industrial sectors where the experience gained from this exercise can be usefully utilized.

Conclusions

The heat pump scheme at the Purley Way Service Centre has provided SEGAS' personnel with an excellent opportunity to gain design, installation and operational experience in a real situation. A number of lessons have been learnt in applying this new technology to the British climatological conditions. The heat pump has been found to achieve a fuel saving of 40% over a complete heating season in comparison with a conventional boiler system. Whilst this figure is very impressive, it does not guarantee an acceptable return on investment against a conventional gas boiler system, in this case of intermittent space heating. Significant penetration in this market will depend upon further development work leading to significant cost reductions and improvement in performance and reliability.

Economic viability improves with larger heating loads, longer hours of operation and reduction in the temperature rise required between the heat source and sink. Some of these situations already exist in the commercial and industrial sectors. Experience gained during this exercise will ensure a better informed and responsible approach in the application of this new technology. In the meantime, the Purley Way heat pump system is being used to explore alternative design and control options for optimizing fuel economy.

Acknowledgements

The authors wish to thank south-eastern region of British Gas for permission to publish this paper. They are indebted to many colleagues, in particular P. Durbin, Technical Consultancy Officer and P. Archer, Senior Scientist, for their enthusiastic approach to the system design, monitoring and development work. Contribution of Miss V. Saksena in analysing the vast quantities of data is also gratefully acknowledged.

Appendix I

Brief details of heating system at Purley Way Service Centre

Heat Pump

Manufacturer:	Bauer Warmepumpen, West Germany
Model:	GLW58
Type:	Air-to-water
Heat output:	64 kW at 1800 r/min at 5°C ambient air and 39 kW at 1100 r/min at 5°C ambient air

Water flow temperature:	65°C
Water return temperature:	52°C maximum
Noise rating:	43 dBA at 10 m
Dimensions:	1520 mm × 855 mm × 1417 mm high (heat pump)
	3050 mm × 565 mm × 1040 mm high (evaporator)
Dry weight:	720 kg (heat pump)
	194 kg (evaporator)
	2 × 178 kg (thermal store)

Supplementary boiler

Manufacturer:	Ideal Standard
Model:	Concord C330
Output:	96 kW
Dimensions:	1130 mm × 556 mm × 1000 mm high
Weight:	670 kg

Service hot water

Manufacturer:	Prometheus Appliances Ltd, Banbury, Oxon
Model:	Lochinvar 75G3
Gas input:	21 kW
Storage:	280 litres

Heating circuit

Type:	Sealed, pressurized to 1 bar
External design temperature:	−1°C
Internal design temperature:	19°C and 16.5°C in store
Flow water temperature:	65°C
Return water temperature:	45°C
Heat emitters:	Stelrad steel extended surface type

Discussion

R. Gluckman (W. S. Atkins & Partners)   A corollary to having the boiler and engine in separate rooms is the siting of the engine air intake. The engine driving a heat pump installation in Lincolnshire has worn to an incredible extent because of a refrigerant leak in the engine room. Refrigerant was ingested into the engine air intake and the phosgene and acids, formed as R12 breaks down, ruined the engine.

What was the evaporator fin spacing used at Purley Way and were there regular frost problems? Dr Gregory (Chapter 4) intimated that with wider fin spacing on evaporators there is less frost.

B. Dann (South East Gas)   An appliance balanced flue terminal was utilized to supply fresh air to the engine carburettor using a flexible pipe connection. We have been aware for some time of the problems associated with refrigerant and gas burning appliances, particularly in hairdressers where the aerosol propellants have a detrimental effect on flues and combustion chambers.

The evaporator at Purley Way was a standard heat exchange unit to minimize costs and the spacings are approximately 2.4 mm. In the 1983/84 winter defrosting occurred very briefly on one occasion early one morning.

Dr V. Sharma (South East Gas)   It is not surprising the evaporator did not need too much defrosting for in 1983/84 there was only one day when the average temperature was less than 3°C and only a few hours when the temperature was sub-zero.

S. F. Barker (Property Services Agency)   Could the authors clarify the position of the buffer tank required by the manufacturers to minimize excessive cycling. In a chilled water system a buffer tank is usually located in the return – a thermal storage tank goes elsewhere and is fitted for quite a different purpose.

B. Dann (South East Gas)   The buffer store is installed in flow and return pipework between the heat pump and the main building circulating pumps (see schematic Figure 5.4). The prime objective of the store is to minimize the frequency of the operation of the heat pump in mild weather conditions and also, according to the suppliers, to damp out vibrations between the heat pump and building pipework. We have questioned the need for this storage in the system particularly in view of the heat losses from it; however, because of the difficulty of isolating the two units, it has not been possible to run the system without storage in the heating circuit.

6 Practical economics of heat pumps in buildings

R. B. Watson and P. M. Robbins

Introduction

With the recent emphasis on the development of heat pumps there has been substantial interest in realizing their full potential in the building construction industry. At the first Heat Pumps for Buildings Conference in Nottingham in 1983, W. M. Currie of ETSU was circumspect in his evaluation of the growth of the heat pump market in the UK. In Germany [1], which appeared to be offering a bright future for the innovation of heat pumps in 1981, there has been a dramatic decline in their home market, the worst contraction occurring in the market for electric heat pumps for space heating.

It must therefore be apposite to turn to economic criteria in order to assess the future potential of the heat pump, for it is not just sophisticated technical innovation which will encourage users to accept heat pumps as an integral part of their environmental services plant, but commercial viability.

If the application of heat pumps for buildings in the UK is to be mainly in space and water heating, then the three main components of financial evaluation for this sector require to be examined: capital cost, running costs and trends in fuel costs. These have to be compared with their more conventional counterparts and a favourable comparison produced, if heat pumps are to be regarded as a true viable alternative.

Not every area to which heat pumps can be applied will produce such a result. The areas which have proved to be worthy of consideration to date are sheltered housing, certain types of specialist medical units, hotel and office accommodation, swimming pools and retail store outlets. In each of those areas there is a regular demand for heat over a relatively continuous period, or there are complementary features in the form of heating or cooling requirements which justify such consideration. These features are able to make a contribution to the overall economic viability of the scheme and the concomitant conservation of energy resources.

Capital cost

Whether the design features incorporated in the heat pump installation are simplistic or sophisticated, whether the energy source is unlimited or restricted, or whether the performance and efficiency factors appear attractive or marginal, the basis of the calculations for economic viability will depend on the initial capital cost of the heat pump itself.

The source of these initial costs is of course the manufacturer, but each manufacturer will have incorporated different design features which creates difficulties when directly comparable figures are required in order to establish costs for comparison.

Table 6.1 sets out the comparative costs for electric heat pumps for the four media, air/air, water/air, air/water and water/water, based on four selected kW outputs and calculated from quotations obtained from a minimum of six sources. Quotations from manufacturers which were demonstrably outside the norm for the size and type of heat pump in question have been discounted to present a more balanced view.

Table 6.1   *Heat pump supply only cost in £/kW 1984*

Heat pump type	£/kW			
	10 kW	*30 kW*	*70 kW*	*100 kW*
Electric				
Air/air	165	185	195	155
Air/water	210	220	170	190
Water/air	120	100	—	—
Water/water	190	130	110	95

The prices set out in Table 6.1 are expressed in £/kW, based on an average coefficient of performance for each size, and this explains the apparent reduction in cost for the heat pumps with the higher outputs.

The equivalent prices for gas heat pumps are not so readily obtainable for comparison at the lower output range and commercially these are generally obtainable only for more industrial applications.

Table 6.2 sets out the equivalent costs to Table 6.1 for the supply and installation of those same heat pump types, to enable comparisons to be made with tender prices for plants in conventional installations.

Table 6.2   *As installed costs in £/kW*

Heat pump type	*£/kW*			
	10 kW	*30 kW*	*70 kW*	*100 kW*
Electric				
Air/air	185	205	215	170
Air/water	235	245	190	210
Water/air	140	115	—	—
Water/water	215	145	125	105

In establishing comparisons with conventional installations however, the costs of the heating plant alone are insufficient if the true additional cost is to be ascertained for a scheme incorporating a heat pump solution.

It is therefore necessary to analyse the costs of the mechanical and electrical engineering services for the various types of buildings identified earlier and if possible to isolate the costs of the heat pump in each case.

Table 6.3 sets out two such types, each type relating to a typical area commensurate with that type of building.

The areas chosen have been rationalized on the basis of 2000 m² for the retail outlet and 7000 m² for a typical office block. The resultant costs indicate typical figures for that particular installation and represent average costs which would be expected to apply in each case.

To make a realistic comparison, the design for a project would have to be produced both for a scheme with and a scheme without a heat pump, within the same building and priced under the same tender conditions.

To find a suitable scheme for such a comparison to be made is therefore extremely difficult, but Table 6.4 [2] sets out the result of an exercise where such conditions were met. This scheme related to a swimming pool which was designed originally for gas boiler heating but subsequently altered to include heat recovery from heat pumps.

In this particular project the swimming pool was approximately

Table 6.3 *Cost breakdown*

BCIS Ref.	Elements	Department stores Area 2000 m² £/m²	Department stores Area 2000 m² £	Offices Area 7000 m² £/m²	Offices Area 7000 m² £
5D	Water Installations	3.70	7,400	7.43	52,000
5E	Heat Source	5.60	11,200	3.20	22,400
	Heat Pumps	39.90	79,800	9.00	63,000
5F	Heating Installation	8.15	16,300	20.29	142,000
	Chilled & Cooling Water	—	—	10.17	71,200
	Supply & Extract Ductwork	22.30	44,600	24.20	169,400
	Air Conditioning Plant	Inc.	Inc.	52.64	368,500
	Cooling Plant	Inc.	Inc.	10.18	71,300
	Automatic Controls	5.85	11,700	13.86	97,000
5H	Main Switchgear	5.00	10,000	9.51	66,600
	Sub-Mains & Boards	6.75	13,500	8.50	59,500
	Power Supplies	6.05	12,100	18.26	127,800
	Lighting Installation	16.00	32,000	19.41	135,900
5I	Gas Installation	1.10	2,200	0.07	500
5K	Protective Installations	19.50	39,000	4.13	28,900
5L	Communication Installations	0.55	1,100	6.99	48,900
	TOTALS	140.45	280,900	217.84	1,524,900

325 m² in area, with a training pool of approximately 45 m². Seven gas boilers were provided to heat shower water, the radiators in the offices and the supply air, together with the heating required for the pool water itself.

Run-around coils were provided to operate in conjunction with two heat pumps to recover heat from exhaust air and to direct it to the supply air or pool water as necessary. During any high demand periods,

Table 6.4 *Comparative costs*

BCIS/CI/SfB Refs.	Elements	With heat pump		Without heat pump	
		Cost £/m²	*Cost of element in £*	*Cost £/m²*	*Cost of element in £*
5D/53	Water supply	2.69	990	2.69	990
5E/55	Heating plant	90.61	33,345	90.61	33,345
	Heat pump	114.08	41,982	—	—
5F/57	Ventilation installation	161.00	59,420	161.00	59,420
5H/61	Electrical installations	39.35	14,479	38.21	14,063
5M/53	Filtration	91.64	33,724	88.91	32,724
5M/53	Water treatment and special services	23.00	8,559	23.00	8,559
		522.37	192,498	404.42	149,201

the gas boilers were controlled to cut in to cover any shortfall. One heat pump operated for the pool hall and the other for the changing rooms' ventilation.

Examination of Table 6.4, however, reveals that no apparent decision was made to reduce the heat source requirements when the heat pump was incorporated into the scheme and this apparent unwillingness to recognize fully the contribution made by the heat pump is by no means an isolated example. Several projects examined betrayed this apparent lack of confidence on the part of the designer and in at least one case, the resultant unnecessary increase in capital cost led to the exclusion of the heat pump in the final scheme.

Identification of cost factors

Cost factors which influence or contribute to the capital cost of the installation of heat pumps are seldom susceptible to independent analysis.

Such factors include the assessment of the coefficient of performance, the adoption or otherwise of the bivalent system and the selection of air/air or other alternative types of heat pump, the degree of sophistication of the automatic controls, the general insulation standards of the building, the disposition of the plant within the building, the methods of air and water distribution, the degree of back-up facilities for standby, the availability of or constraints to available capital expenditure, the choice of refrigerant and the availability of compatible plant operating in parallel.

Energy conservation

Once the additional capital cost for the incorporation of the heat pump into the project has been identified, the factors which influence or have contributed to this cost should also be identified and analysed separately where possible.

Due to the variety of heat recovery systems which are installed to operate in conjunction with heat pumps, the effect on other elements of incorporating the heat pump will vary considerably. There are many parameters in adopting an energy saving scheme of which the heat pump is only one.

Heat recovery systems can include obtaining the heat from exhaust air, flue gases, food or other refrigeration processes, cooling systems, natural sources, etc.

The economic viability of the heat pump will therefore depend on the context into which it is installed. Heat pumps may work in conjunction with dehumidification processes associated with swimming pools; they may depend on the availability of a suitable heat sink as with groundwater heat pumps; they may operate more effectively with thermal storage facilities; they may be operated in conjunction with compressors installed for separate refrigeration processes; or they may require additional boost from a solar heat collector.

The economic viability of the heat pump can therefore depend upon the installation of both associated heat recovery systems and the features of compatible plant installed for other purposes. The identification of these other factors could lead to the recognition of a reduction of the additional capital cost of the heat pump when isolated in this way.

Running costs

Running costs can be divided into fuel costs, maintenance costs and, eventually, replacement costs. Generally, the replacement costs would not form part of the overall economic calculation, but items of mechanical plant usually have a sufficiently short life that occasionally, these costs cannot be completely ignored.

So far as heat pumps are concerned, one of the primary advantages of installing a heat pump stems from the direct savings in fuel costs. Unless, therefore, the total capital cost of the heat pump selected is above certain limits, the rewards related to the capital investment will be derived from cheaper energy bills. In many cases, the future benefits from the successful employment of the heat pump will be sufficient to warrant the initial capital investment.

As with the exercises required to identify additional capital costs, it is also necessary to identify the additional running costs, calculated from the cost of energy consumed in those buildings with heat pumps, when compared with the corresponding cost of energy consumption relating to those buildings using alternative design solutions.

The calculation of these costs should where possible include all ancillary fuel costs of the plant types under consideration. For example, energy consumption of the fans associated with an air source evaporator may be as much as 10% of the compressor input, with the consequential increase in fuel costs [3].

Figures available for the total cost of energy for two common building types, each with different mechanical installation design alternatives and each with heat pumps, gas or oil as a heat source, are set out in Table 6.5. The figures represent average costs for two of the more common types of installation for each building.

In comparison, the recorded electrical consumption figures for a High Street shop over a 12-month period in 1977–1978 are set out in

Table 6.5   *Average energy costs for two building types*

Building	Total building energy cost £/m² per year		
	Heat pump	*Gas*	*Oil*
1	5.29	5.37	6.03
2	4.89	4.93	5.80

Gas oil	– 18.5p/litre
Gas	– 35.2p/therm
Electricity	– 5p/unit
COP heat pump	– 3.5

Table 6.6 *Energy costs for air to air heat pump in a high street shop*

Electricity	Consumption W/m^2	Cost $£/m^2$
Heat pump heating	15.68	0.78
Heat pump cooling	11.59	0.58
Heating & ventilating general	28.71	1.43
Lights & small power	113.38	5.66
Totals	169.36	8.45

Table 6.6 [4]. This shows a breakdown in terms of £/m², as well as the electricity consumed over the four main elements supplied with electrical power. The electricity is based on a charge of 5p/unit to reflect todays electricity prices.

As an alternative view of fuel costs, an exercise has been carried out which gives figures for an assumed heating load of 29.31 kW, with a night setback of 12 hours. The results and relevant fuel price assumptions are laid down in Table 6.7 [5]. The intention of this table is not to relate the costs back to area, but to establish a basis for comparison. However, a recent exercise for a first floor office, with an average level of insulation and approximately 325 m² in area, had a calculated heat loss of 29.31 kW.

Table 6.7 *Fuel costs for 29.31 kW with night set back of 12 hours*

System	Fuel price	Annual heating cost £
Direct electric	6.5p/kWh	4035.01
Oil	1.10p/gal	2312.88
Gas	35p/therm	1042.00
Packaged air/air heat pump	Electric 6.5p/kWh COP 2.6	1549.09
Split systems air/air heat pump	Electric 6.5p/kWh COP 4.16	969.28

Maintenance costs

The subject of reliability appears at present to be a moot point in respect of heat pump installations and the degree of maintenance required. It is generally accepted, however, that a heat pump with its associated controls is more complex than the conventional alternatives and therefore requires a greater degree of supervision to ensure efficient operation, if the desired pay-back period is to be attained.

A number of examples have been found of unsuccessful start-up, unreliable operation or an absence of proper maintenance, leading to a requirement for back-up fossil fuel heating for excessive periods. In such situations, where the heat pump cannot regain energy lost when back on-line, there is a corresponding extension to the pay-back period. Fortunately, such projects are counter-balanced by those which operate successfully to give a satisfactory performance and a corresponding achievement of forecast savings.

Irrespective of the design parameters selected and the running costs forecast, actual running costs for comparative analysis continue to remain elusive. The efficiency of the system may be impaired by initial teething problems or inadequate maintenance, the fuel sources may be shared with other equipment and prospective fuel costs will be subjected to irrational increases arising from both economic and political factors. In addition, there is the human factor, which allows energy to be wasted, energy conservation to go unrecognized and budget systems which fail to allow running costs to be associated directly with initial capital expenditure.

Nonetheless, suitable data have become available from certain diverse sources, although it is not always easy to draw reliable conclusions in every case.

It has been suggested that by examining figures obtained from the refrigeration industry, a proper allowance for maintenance costs for heat pump installations, for both water source and air source heat pumps, would be 5% of capital cost per annum, with a further allowance of 7½% of capital cost for maintenance of a water source well where appropriate [6].

In terms of total cost per annum, the maintenance costs for 2700 m² of sales area in a supermarket in 1982 produced the following figures.

Central gas fired plant   £340 per annum
Rooftop packaged unit with   £450 per annum
   gas fired boiler plant

Air to air heat pump   £550 per annum
All electric installation   £240 per annum

Maintenance costs can be expressed as a proportion to shaft power and, based on an exercise in 1980 for gas engine heat pumps [7], calculated on 0.2p/kWh of shaft power, this would produce the computation:

$$\text{Maintenance in p/kWh} = \frac{0.2}{COP_H + 2.115}$$

As a final basis for comparison, the comparative costs of different heating systems were assessed recently by the Electricity Council. Table 6.8 indicates the approximate figures of maintenance costs per annum on the basis of £/m² for seven different systems.

Table 6.8   *Comparative costs of heating systems*

Heat source type	Maintenance £/m² per annum
Storage heating	0.25
Storage direct	0.25
Central water storage	0.80
Direct electricity	0.25
Gas	1.0
Oil	1.2
Heat pump	2.0

Heat pump control systems

The control systems installed with a heat pump installation will have a direct effect on the efficiency of the installation, together with its running costs. If the heat pump is to operate at optimum performance, the control systems have not only to be installed correctly, but proper technical information must be provided and followed by the operator.

The additional costs of installing suitable control systems require to be identified separately. Less complex energy conscious schemes may not require the same capital outlay, but may utilize greater manpower requirements in their maintenance. A more expensive controls installation, once it has been properly installed and set into operation, may reduce running costs disproportionately.

Evaluation of economic viability

The economic viability of a heat pump can be evaluated in terms of years, in the form of the pay-back period required for the additional capital cost of the energy saving measure adopted, to be offset by the estimated savings at present-day prices on an annual basis. This ratio has then to be modified by taking into account the percentage adjustment for the annual premium in energy costs. In producing an acceptable figure for this ratio, it is generally recognized that energy costs would have to rise disproportionately in relation to general annual inflation levels before the additional capital cost of incorporating heat pumps in a project would become a viable proposition. According to the Electricity and Gas Councils, the immediate future for the next 5 years does not indicate that either electricity or gas will increase in cost above the general level of inflation. The viability is further reduced when one considers the relatively low number of years before the equipment will require to be renewed.

However, energy costs are influenced by a number of external factors such as the strength of the dollar and its effect on oil prices, the eventual increasing difficulty of extracting gas and oil from the North Sea, industrial action and its effect on coal prices, and the strength of sterling in world markets. The combination of these influences is likely to make it difficult for energy costs to be contained indefinitely within the general inflation rate.

To give some indication of the relationship between capital and running costs, reference can be made to Table 6.9 which indicates a cost comparison for capital and running costs between conventional and heat pump treatments, based on a school swimming pool [8]. Prices relate to mid-1981 and running costs were estimated using the methods adopted by the Electricity Council.

The additional cost of the heat pump, coupled with the corresponding saving in running costs, will produce a simple pay-back period of

Table 6.9 *Capital and running costs for school swimming pool with conventional and heat pumping treatments*

System	Capital cost	Running costs pa
Conventional	£35,385	£19,321
Heat pump	£85,780	£11,442

6.4 years. The figures relate to a smallish pool of 213 m² and larger pools of, say 320 m², are inevitably likely to yield a small payback period. It has been claimed that such a period might be in the region of 2–3 years, but this makes no allowance for the cost of the capital involved in investment, as re-payments related to the initial capital are excluded from such a calculation.

Relationship between capital and running costs

It is important to relate an increase in capital expenditure to a saving in running costs. In some cases, the saving in running costs will not justify the proposed increase in capital expenditure unless certain economic factors are taken into account. The establishment of the relationship between capital and running costs depends upon capital cost calculated at the present time being compared with running costs to be expended in the future, calculated in equivalent present-day terms.

In order to evaluate running costs in terms of the above, account has to be taken of the present value of the pound for running costs forecast for succeeding years; the cost of funding capital expenditure at appropriate interest rates; techniques for assessing comparative costs such as discounted cash flow; fiscal policy regarding allowances on plant and machinery; levels of corporation and other taxes; and the life of individual items of plant and the residual value at the end of their useful life.

In order to make an appraisal on a rule-of-thumb basis to take into account these figures, it is perhaps worth stating in broad terms that the additional capital expenditure on engineering plant of, say, £10,000 would require a reduction in running costs of approximately £950 per annum [9].

The above calculation assumes that capital would be raised on the basis of a mortgage and would require to be re-paid over a twenty-year period at a 12½% rate of interest. Re-payment would be alleviated by relief on corporation tax, which would otherwise be levied at the rate of 50%. For tax purposes, engineering plant is often regarded as fixtures and fittings, rather than as plant and machinery, and may have an assumed life of 20 years. Running costs are calculated on the basis of a 5% rate of inflation per annum.

The economics of comfort standards

Irrespective of the precision or lack of precision by which the economic viability has been calculated, there are a number of non-economic viability factors which may impinge upon these figures or influence their acceptance on a basis which would otherwise lead to their rejection.

A swimming pool offering a pleasant environment will attract more visitors. An office or factory with good environmental conditions will contribute to better production and stability of staff levels. A higher quality of environment can add to prestige, a stable environment in a retail outlet will contribute to greater turnover. The adoption of an innovation in the forefront of technical development can be used as an advertising feature. Against a background of factors such as these, simple economics may not form the sole factor for basing the decisions to incorporate a heat pump into a design scheme.

Conclusions

In examining the capital cost of heat pumps, it would appear that the contribution to the overall design has often been under-estimated and the plant installed, as a source of both heating and cooling, has not always been reduced in capacity to take fully into account the contribution made by the heat pump.

In establishing the running costs, the energy requirements of the heat pump have seldom been isolated, with the result that the savings in fuel costs have often been supplemented by the savings accruing from other energy-saving features of the project.

The combination of the above two factors has inevitably led to a distortion in the true economic viability of the heat pump. On the one hand, the apparent additional capital cost has been exaggerated, while this has been offset in many instances by a more favourable set of calculations from anticipated running costs.

Where heat pumps have been installed in conjunction with air conditioning systems or swimming pools, favourable results have usually been forthcoming, although in many cases figures affected by accountancy techniques will often produce a less favourable picture.

References

1 Bad Year for Heat Pumps, *Building Services and Environmental Engineering*, July 1984.
2 By courtesy of P. Faulkner of Raymond Allington Partnership.
3 Hogarth M. L. & Pickup G. A. (1983) *The Role of Gas-fuelled Heat Pumps for Space Heating*. From Heat Pumps for Buildings Conference 1983.
4 *The Energy Consumption of a Heat Pump in a High Street Shop.* The Electricity Council, January 1979, Ref ECR/R1213.
5 Design Installation Services (Cheltenham) Limited.
6 *Ground-water Source Heat Pumps for Space Heating in the UK.* W. S. Atkins & Partners, under contract to ETSU.
7 Masters J., Pearson J. & Read M. A. (1980) Opportunities for Gas Engine-driven Heat Pumps in the Industrial and Commercial Markets. British Gas.
8 By courtesy of Haden Young.
9 Watson R. B. (1980) Capital and running costs of air conditioning systems. In *Air Conditioning and Energy Conservation*, Editor Sherratt A. F. C., 1980.

Discussion

J. Mundy (Cybus Energy Services Ltd)   There seems an enormous discrepancy between the packaged air to air heat pump and the split system in annual heating costs shown in Table 6.7. One is £1549.00 and the other is £969.00. Is there some reason for that? If direct electric heating (62000 kWh) is taken as the heat load the COP implied by the cost figures for the air/air heat pump system is over 4.0, which appears high.

P. Robbins (Northcroft Neighbour & Nicholson)   These were given for a specific heat load or project. They are based on two particular heat pumps and from the enquiries to our source we were led to believe that the difference is due to the design configurations of the actual heat pumps, the principal difference being the compressors; one unit had only one compressor, the other had two which allowed better part capacity operation and hence increased efficiency. We would assume therefore that it is very important to carry out the calculations and get the right unit for a particular building. The COPs for the two systems are: (a) packaged ir/air heat pumps, COP 2.25; (b) split system air/air heat pumps, COP 3.40. The COP is calculated by division of heat pump output by heat pump input at an ambient temperature of $-1°C$.

R. A. Harrison (British Home Stores plc)   Mr Watson acknowledged that high street stores have been among the leaders in the use of heat pumps. However, the importance of space was not mentioned. A major advantage of a heat pump is that it can be roof mounted and have an integral cladding, releasing very important space in the store itself.

Those responsible for selecting the most suitable plant and system design for a building have available to them various computer programs operated by the fuel interests such as the Electricity Council and the local Gas Board. Other professional bodies are also devoting considerable resource to developing independent programs. Would it not be better if the Department of Energy, or more specifically the Energy Efficiency Office, were to select and appoint one central body to provide this service? This central body could then provide independent advice as to which was the most suitable configuration of plant for a particular building. As a condition the client could be asked to provide subsequent actual performance figures against target, we would then have a learning curve in the master computer program from which others could benefit.

G. W. Aylott (The Electricity Council)   The Electricity Council analysis program BEEP may be used without charge and is still regularly and widely used. However, it is 10 years old and we are working on improvement, and particularly in user friendliness and interactive operation.

M. D. Terry (North Thames Gas)   What real relevance are 20-year mortgage pay-backs when the equipment concerned will probably have a life of much less than this term?

R. B. Watson (Davis Belfield & Everest)   People presume that capital will be available for a building project when this is not necessarily so. In many cases the initial capital has to be borrowed and under those circumstances payment in the form of interest has to be made to enable that money to be available. It is not unrealistic to use a 20-year period as a basis of calculation of this interest, although I do take your point that mechanical services plant in particular may have a life expectancy of only half that period. Nonetheless, the total sum of money to fund the capital requirements at the beginning of a project still requires to be raised and the interest charges on that amount spread over a 20-year period.

7 Experience of some large-scale heat pump installations in Germany

J. Paul

Summary

It is some ten years since the first large-scale heat pumps have been installed, after the first oil shock in 1973. The introduction of large-scale heat pumps generated a lot of new ideas; this paper concentrates on remarks and findings, which can be generalized. Essentially, it is not the heat pump which causes troubles, but peripheral installations, operation and design. To obtain the best results, and avoid the majority of problems, we should concentrate on simple heat pumps.

Heat pump installations and heating system

Generally speaking, most of the heat pump installations are much too complicated. It seems to be a major feature of designers and contracting companies to develop fancy systems and produce 'sophisticated' equipment. Heat pump installations mostly consist of the heat pump itself, boilers and the heating system. The heat source may be an additional feature which can easily become excessively complicated. Fancy plants neither meet the customers' needs, nor are they an enjoyable reference for the contractor and consultant.

'Sophisticated' systems can be recognized by an enormous network of controls, causing faults and errors in the plant and reducing the economy of its operation due to failures and permanent trouble shooting.

The principal recommendation for multiple heat pump systems is to provide separate refrigerant circuits for each unit, linking together the units of the water side, but avoiding connection in the refrigerant side. The heat pump itself is safe and reliable if the circuits are kept simple. We have changed many complex designs from consultants and produced 'autonomous units' which have the simple coupling on the water side and are easily regulated by individual controls.

It seems to be a general misconception that heat pumps of high capacity are especially economical and effective. Mostly, the chosen capacity of the heat pump is too high in respect of the heating demand; this results in a lower annual running time than would normally be expected.

The ratio of boiler to heat pump capacity is in many cases unfavourable, especially if this ratio has been derived from the installed boiler capacity without determining the real heating demand.

The design of heating systems which are interacting with the heat pump does not reflect, in the majority of cases, the characteristics of heat pumps. This refers not only to *temperature levels* and *temperature differences* of the hot water heating system, but also to the common habit that two different companies are involved – one for the heat pump, another for the heating system – signifying not only a lack of communication and information, but also the absence of classified responsibilities.

Often 'mixed circulation' may affect the heat pump. This mixed circulation is water flow from the boiler, which is fed into the heat pump condenser. The reason for this is not only the wrong kind of piping, but also incorrect placement of temperature sensors for control. Experience shows that conventional ways of thinking for heating systems are not usually satisfactory for heat pump installations.

The actual *water temperatures* of a hot water heating system are usually lower than pre-calculated with respect to flow water, and higher than pre-calculated with respect to return water temperature, resulting in insufficient temperature differences in the heating system. Under part-load conditions the condensing temperature could, therefore, be too high. The safety cut-off of the heat pump, (over-pressure) or a lower COP, is the result of erroneous operation of the heating system.

The following example illustrates how theoretical ideas may be totally different from those applied in practice.

The design return water temperature was 37°C, the actual conditions showed a mean return temperature of 51°C. If the design flow water temperature of 60.5°C were to be reached, the design temperature difference of the heating system would be 23.5 K. The actual temperature difference is only 9.5 K; this means that the design temperature difference is 2.5 times higher. This results in premature cut-offs of the heat pump or in part-load operation, both contravening the basic idea behind the decision to install the heat pump.

Another basic mistake may be found in the choice of *circulating pumps* which are either too small or too large. If the water flow in the circulating system is insufficient, the temperature difference in the condenser of the heat pump becomes too high; the heat pump is either shut off or running under part-load conditions. If the capacity of the circulating system is too high, the flow water temperature may be too low, the boiler could take over too early and thereby reduce the annual running hours.

In the case of combustion engine drive, the *high and low temperature circuits* are often managed in the wrong way. Either uneconomic mixing of two different temperature levels is achieved or – if the design of heating systems does not meet the actual conditions – even a cooling tower for high temperature management must be used. In both cases it is found that useful energy is downgraded from the high temperature side, an offence to basic thermodynamic laws. Especially, when considering the use of large-scale heat pumps, *thermal storage* has not been taken into account, or at least not sufficiently. If thermal storage is insufficient in the building, thermal storage tanks are too small or omitted altogether, the heat pump system becomes nervous, that is, the system is being switched on and off too frequently. If the building, or heating system itself, is not capable of absorbing the heat pump capacity for a period of at least 20 minutes, storage should always be contemplated as a viable alternative.

If a *batch storage tank* is applied, the hot/cold layer can be used for control of the heat pump, providing the movement of this layer is monitored. Taking into account the position of the layer, the heat pump can be efficiently controlled with respect to part-load operation, and annual running hours can be increased without the risk of premature cut-offs.

As the authorities in Germany enforce strong regulations on pressure vessels, expensive-to-satisfy steam vessel regulations must be applied if possible water flows from boilers above a temperature of 100°C have been overlooked. It is impossible to utilize installed pressure vessels if this is not rectified immediately. Therefore, hot water systems must always be checked to see whether pressure vessels are influenced by the boiler, and how this influence is likely to affect safety regulations.

Heat source related matters

Possibly due to wishful thinking during the planning process, or over-optimistic values, the heat source temperatures are often lower in practice than has been expected during the design of the system. The result of these insufficient temperatures is that the heat source is cut off prematurely, reducing the COP and thereby annual running hours. The risk of freezing or growing ice on the heat exchangers may be a possible danger, especially for water as heat source.

If *river* or *lake water* is utilized, this may carry too much dirt, either fine particles or larger kind of materials, such as wood and plants. Shell-and-tube evaporators may therefore become blocked. Excessive costs for cleaning either the entering water or the blocked evaporator must be foreseen. Preventative measures could be taken by using fine filters with reverse operation for cleaning, air showers where compressed air is blown into a blocked evaporator, or special measures for design to extract water without risking too much dirt or other particles. The best solution is to go for an evaporator which is not sensitive in this respect to dirty water and can be cleaned easily.

With *air as a heat source*, the air flow pattern on evaporators is often very irregular, resulting in non-uniform frosting of heat transfer surfaces. Frequent defrosting cycles are needed, reducing the economic viability. Unsatisfactory defrosting, commonly linked with poor operation of expansion valves, is present. The expansion valves are sometimes responsible for non-uniform refrigerant flows into the evaporator, and also a source of noise if the capacity of the thermostatic expansion valve is wrongly chosen.

Experience related to refrigerant circuits

Generally speaking, the refrigerant circuit itself is not the real problem of heat pump operation, but moderate peripheral installations, often resulting in problems with the refrigerant circuit itself.

As stated previously, the chosen capacity of the heat pump is often too high. In this case the *part-load efficiency* is very important, as well as the choice of compressors, with respect to type and capacity mix. With reference to winter and summer operation, as well as to frequent part-load conditions, *type and capacity mix* of the compressors should meet economic requirements to the utmost extent.

With respect to oil in heat pump plants, *oil return systems* do not always meet the requirements and result in an accumulation of oil in evaporators. The decrease of COP and economy is obtained by reducing the evaporating temperature.

With respect to mineral and synthetic oils, viscosity and solubility for oil and refrigerant, it must be taken into account – especially for screw compressor units – that a detailed know-how of *oil/refrigerant characteristics* is necessary. If liquid refrigerant is allowed to enter, for example, a screw compressor, the separation of oil from refrigerant in the oil separator may become a problem, especially if the solubility of refrigerant in the oil is high. The oil is going to foam up and be transported in excessive quantities into the system, thus reducing efficiency and effective heat exchanger surface.

For large-capacity evaporators, *thermostatic expansion valves* may interfere and cause either unstable operation or result in noise and vibration problems. Also the distribution of refrigerant can suffer, thereby giving too low COP.

For water or brine cooling with shell and tube heat exchangers, special care must be taken with respect to the *flow velocity* of the liquid in dry expansion evaporators. Under part-load conditions, where the water flow may be reduced, there is a risk of laminar flow. The reduction of velocity is especially dangerous if no pressure drops are present in the water or brine; a reduction of 20–30% is the maximum that can be aimed for.

Combustion gas engines as prime movers

Special measures must be taken for *starting* gas engines. Multi-start-up is necessary, allowing for at least three start-ups, due to safety measures needed. The start of the engine without activated ignition is necessary in order to remove residual gas, to prevent unsafe conditions or actual damage to the unit. If a gas engine is cut off, usually a solenoid valve in the gas in-take pipe is closed, the engine dies from lack of gas and removes most of the gas remaining in the pipes. The procedure described below is recommended in order to prevent danger and also to comply with safety regulations.

Figure 7.1 shows the general lay-out consisting of two solenoid valves which allow for monitoring gas leakages. If a gas engine has to be cut-off, the first solenoid valve situated next to the engine should be

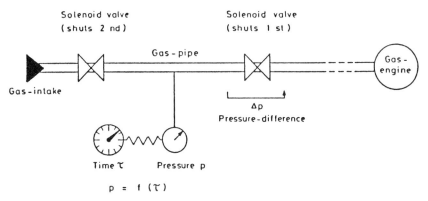

Figure 7.1 *Safety measures for gas engines.*

closed, causing the engine to die due to lack of fuel. The second solenoid valve should be closed some seconds after the first one. Both valves enclose a certain volume of gas, which can be measured by a pressure gauge. As there is a pressure difference between the enclosed gas volume and the pipe leading to the gas engine, the leakage of the solenoid valve will reduce the pressure in the enclosed gas volume. By monitoring the pressure, and taking into account the time in which the pressure drop may occur, gaseous leakages from the gas in-take to the gas engine can be detected. With these two solenoid valves, not only is the utmost safety ensured but also the level of danger is reduced, due to the start-up of the gas engine without ignition.

If the *combustion temperature* of the gas engine is too high, the sparking plugs wear too quickly and may become severely damaged. For example, sparking plugs were ejected from the engine because they could not withstand high exhaust gas temperatures. In this case gas from the cylinder, without the spark plug, escapes to the machine room. Therefore, the exhaust gas temperature must be adjusted very carefully.

If the *exhaust temperature* after the exhaust heat exchanger is too high, thermal losses occur which reduce the energy output. If the exhaust temperature after the exhaust heat exchanger is too low, and if the condensate is not drained off in the proper manner, or if stainless steel is not used, there is a risk of corrosion of pipes and exhaust equipment.

For the material used in the exhaust system, both stainless steel and plain (black) steel may well be considered. The price level of stainless steel as compared to plain steel is roughly about 7:1. For example, stainless-steel pipe of 150 mm diameter, including fittings and insulation, costs approximately 1000–1200 DM per metre, in plain steel the same pipe would only cost 150–200 DM per metre. Whether plain steel should be applied and the exhaust system replaced more often is purely an economic case. It is, however, always essential to drain-off remaining liquid (condensate) as these remains are concentrated more and more with aggressive components, due to the fact that water is continuously evaporated from this condensate.

General remarks

An old question is who is responsible for the heat pump plant if this plant is commissioned from different companies. There are many bad examples of this and – except for component sales – customers for heat pump plants are always better off if they buy the whole thing from one main contractor, only undertaking joint ventures under special circumstances.

With respect to *noise*, Table 7.1 gives some ideas of noise levels of main components.

Table 7.1 *Noise levels of main components of heat pump plant*

Components	Noise level (approx.)
Reciprocating compressors	84 dB (A)
Screw compressors	92–96 dB (A)
Electric motors	72 dB (A) (air cooled)
Gas engines	96–100 dB (A)

Special attention must, therefore, be paid to the placement of the *plant room* within a building. There are many examples of the location of the gas engine-driven heat pump being the last decision taken, and in one case, in a hospital, the plant room had been placed underneath the office of the Chief Medical Officer. It is no wonder that he complained about noise and vibration – just think about the idea of having a gas engine running 3 m below *your* desk!

Vibrations can be compensated for by using rubber compensators in the hydraulic system. Vibration dampers for refrigerant circuits are more difficult and less efficient as, due to the high pressure, they need special material and strength, which reduces the ability to compensate for vibrations.

In our experience, *foundations* are not necessary and their importance is mostly over-estimated. If these foundations can be left out, it will not only save money, but also make transportation and installation of the system easier as no special cranes and fork lifts, etc. are needed.

If heat pumps were bought according not only to their price, but also to their specific qualities, many better installations could be found. Going for *low-speed* compressors and engines increases the life-time and the efficiency of reciprocating compressors, and reduces considerably the noise of screw compressor units. It is a bad habit in West Germany, and probably in many other countries, that public customers, such as municipalities, have different sources of money to fund heat pumps. For instance, the Ministry of Finance pays for the *investment*, and the appropriate ministry under whose roof the heat pump is running pays the *operating costs*. An example may be the application of heat pumps in universities or schools where the Ministry of Finance pays for the investment, and the Ministry of Education pays for the operation. There is no motivation for the investor to make an installation more economical in the long run, by investing more money initially and reducing operating costs in terms of fuel and service, as the same institution will not pay the consequences.

Another reason for unsatisfactory operation was to a large extent the *electric cabinet*, which includes the necessary control of the heat pump. Many companies making these electric cabinets are not capable of providing adequate service from the very beginning. Expensive and lasting efforts to compensate for this are often necessary. In one case, for example, the wiring for switching off the system due to lack of oil in the compressor, was completely forgotten and during test runs the gas engine did not know that it had to switch off. This signal simply did not reach the engine. The result was complete destruction of the compressor after a couple of hours.

The final remark refers to the *operation and maintenance* of heat pumps. General experience shows that customers' operating personnel are neither willing, nor capable in most cases, of providing for proper operation. The outcome of this is that the service contractor has often been called in order to find out that a circulating pump has not been switched on, or that cooling water of the combustion engine was just missing. In one case the operator noted carefully for weeks how the cooling water level was going down and later called maintenance personnel in the night as the gas engine had switched off due to lack of cooling water. Sometimes those people do not get any message about the operation. Unsatisfactory results come mainly from peripheral problems and part of these problems are the personal capabilities of the customer.

Experience, as illustrated by this paper, has taught us how to make heat pumps more simple, more effective and less risky now and in the future. After 10 years of experience, they have become part of our professional know-how.

Discussion

R. Gluckman (W. S. Atkins & Partners)   Many of Dr Paul's points mirror my own views. Considering esoteric 'university' type techniques such as non-azeotropic mixtures, Lorentz cycles, etc. is an over-complication. There are an incredible number of simple things that can be done to the refrigeration cycle to improve heat pump performance. Perhaps it reflects Dr Paul's involvement in conventional refrigeration that he sees the advantages to be gained by this attitude. Heat pump COP is not a fixed number; the engineering of the refrigeration cycle determines it and this therefore must not be ignored.

The capacity of large-scale refrigeration systems does not increase with increase of ambient temperature – it increases with increase in evaporating pressure. Both Mr Dann (Chapter 5) and Dr Paul noted the former effect and Dr Paul explained that because the condensing temperature was also rising the comparison was with different phenonema. This shows the importance of understanding the exact mechanisms in a heat pump cycle; perhaps the difference between predicted and monitored results is that in the monitoring not all variables are taken into account to be able to fully understand the figures.

Dr R. G. H. Watson (Building Research Establishment)   I am surprised that the thermal content of the air even at $-16°C$ only occasionally provides a sufficient heat load in well-insulated German homes to make heat pumps economic. Is it therefore the summer use that is important?

J. Paul (Sabroe & Co.)   At present the German market for domestic heat pumps is dead. This is mainly due to the fact that 2–4 years ago there were 150 manufacturers of heat pumps all trying to earn big money. The less scrupulous produced installations or equipment which were unsatisfactory. Every heat pump installation that did not work spoiled 100 potential heat pumps. The same problem did not occur with large-scale heat pumps, which are more economic anyway.

In Germany there are three climatic zones – with design temperature −12, −15 and −18°C respectively. A heat pump can be installed to operate down to this temperature on the heat source side either as a monovalent system or a bivalent one but with parallel working. Also if the heat pump is installed outside, the German standard required it to be able to withstand a temperature of −15°C. However, it is becoming recognized that to operate an air source heat pump at −15°C is ridiculous and the general method of operation in practice is as part of a bivalent system in which the heat pump is switched off for 6 hours every day, that is, three times for 2 hours each time. A special tariff is available for such an installation and the general operating conditions for the heat pump are around −3°C to +3°C. The utilities can install a small remote control switch so that by a signal on the mains they can switch the heat pump off and on again. There is also incorporated a probability generator so that the heat pump is not switched on immediately but at any time within 5 minutes of the start signal. This avoids instantaneous switching of a large number of heat pumps. The tariff is rather advantageous but the installation must be provided with at least 2 hours storage in the house structure or buffer storage.

A large number of heat pumps have been installed for use with a groundwater heat source but ground water is now very dubious because of ecological and environmental considerations. The recently revised German standard DIN 8901 specifies safety precautions to be taken in using groundwater to avoid refrigerant and oil escaping into the water.

The attitude in general is positive for heat pumps but at the domestic level the pioneer market is now satisfied – the people who buy heat pumps as a signal of prestige now have them – so for the smaller domestic unit the outlook is a little grey.

M. D. Terry (North Thames Gas)　Dr Paul condemned combining the refrigeration circuits between the different heat pumps. What are his views about two evaporators, or even three, associated with one compressor?

Dr J. Paul (Sabroe & Co.)   One German installation was, following a customer requirement, designed to work on two temperature levels with two evaporators and three compressors. Even having the benefit under part-load of the full evaporator surface, the system was less efficient because the temperature levels of the two evaporators were different. Where compressors are working with multiple evaporators it is the lowest temperature which is responsible for the shaft power. The system was completely redesigned, using autonomous units, that is very simple refrigeration techniques, and it was cheaper, more reliable and more efficient.

Cold stores very often have two temperature zones, a freezing tunnel at $-30°C$ (and a handling area at $0-10°C$). The tendency is quite clear now; go away from interlinked circuits toward separate units, and perhaps have one compressor to switch over from one system to the other.

Of course if the two evaporators were at equal temperatures, the same problem does not occur and it is quite common to combine two mass-produced evaporators to supply a duty rather than have a single evaporator tailor-made.

Dr A. F. C. Sherratt (Thames Polytechnic)   Could Dr Paul explain why in Germany heat pump applications in housing have dropped dramatically. Is it that the Government grants have been stopped? In Germany there is available a domestic interruptable supply tariff, there is no such tariff in the UK. The facility for the utility to be able to switch off heat pumps seems an eminently sensible way of load shedding during periods of peak demand, and provided the duration of the disconnection is not too long it would not be noticed by the consumer. The time is ripe for such a tariff and down-the-line switching to be introduced in the UK.

Dr J. Paul (Sabroe & Co.)   The German domestic heat pump market was not killed – it committed suicide. Grants have now been replaced by tax relief which gives better return for medium and higher income groups. I therefore purchased a solar system for my house because with tax relief it was a reasonable proposition. Tax relief is available for solar systems connection costs to district heating systems and heat pumps. However, I do not believe in any technology which needs support to make it run, because it will not run when it is not supported and thereby continue into the future. To give tax relief recognizes the effect on the national economy of an investment in reducing oil consumption and improving the balance of payments. In this circumstance the heat pump capital cost is largely paid by tax relief and the actual cost is not therefore significant and the interruptable tariff provides very favourable operating conditions. Due to the attractive market many companies became involved, and heat pump systems of the wrong design are installed, at the wrong time and with the wrong equipment. The industry committed suicide because people learned their lesson and now stay clear of heat pumps in favour of other heating methods.

D. Sparks (Staefa Control Systems)   From considerable experience with gas engine and electric heat pumps, mainly in agricultural and swimming pool applications, I am appalled first by the poor quality of the maintenance carried out and the quality of the staff responsible for it, and second at the inadequacy of their control systems. Typically swimming pool heat pumps are installed with existing plant. The control of the heat pump is often designed in isolation and not integrated with the rest of the system. Often the bivalent system, the second source of energy available, is kept running just in case the heat pump cannot meet the demand put upon it. This is a total waste of energy. A control system which integrates the heat pump with the bivalent energy source provides a much more energy-efficient arrangement.

8 Air conditioning and heat pumps: a systems and energy review

J. Leary and B. S. Austin

Summary

The thermal characteristics of buildings are examined and related to the characteristics of air conditioning systems, with particular reference to four building examples. Metered energy and performance data is available from research work carried out on the four buildings and is used to formulate the suggested design parameters and design procedures. The general arrangement of heat pump plant to suit the characteristics of the various air conditioning systems and buildings is presented together with a description of sizing and method of control. In each case the objective of the design is to select heat pumps to cater for the summer cooling load and utilize them to the full to provide heating. The plant which also includes supplementary heating would have sufficient heating and cooling capacity available throughout the year to match the load imposed by the system. Energy consumption, energy cost and capital cost comparisons are given for the various systems.

Introduction

The air to water heat pump provides either chilled water or heating but cannot provide both simultaneously. The mode of operation is changed over at the dictates of a room sensor, or external ambient temperature sensor, or preferably by manual control at the beginning and end of the heating season. Where air to water heat pumps are installed in central plant air conditioning systems for the purpose of supplying heating and cooling, the characteristics of the equipment introduce a number of design considerations which are obviously more critical than when applying conventional refrigeration plant and boiler plant that have full capacity available at all times. In buildings with a large number of rooms with different orientations, there is a considerable diversity of requirements ranging from full cooling to partial heating, or partial cooling to full heating, depending upon the time of year. Therefore, for the successful application of central air to water heat pumps a careful evaluation of the heating and cooling requirements of the building is required for the whole year.

The purpose of this chapter is to review the energy cost advantage of air to water heat pumps and to show how air to water heat pumps and the distribution systems can be engineered to suit the different types of building with their associated thermal characteristics.

Free cooling

Installations that have a substantial winter cooling load would require an additional chiller for winter operation. The peak summer cooling load would be provided by a combination of heat pump and chiller and at the beginning of the heating season the heat pump would change over to the heating mode. The amount of supplementary heating would be greater than for installations that did not require the separate chiller.

A further consideration arises in buildings that have a substantial winter cooling load and that is regarding the choice between 'free cooling' and the use of a chiller with a fixed fresh air quantity. In the latter case, in the interests of energy conservation, it would be preferable if the chiller had a heat recovery facility and that the recovered heat were fed into the main heat pump system serving the remainder of the building. In the majority of buildings it can be safely assumed that all the recovered heat will be used during the heating season.

Certain systems require a considerable amount of reheat. For example, induction systems, that have a generous allowance of primary air, often as high as 3.0 litre/m²s of air conditioned area, would require a large reheat load at all times; 3.0 litre/m²s is approximately four air changes per hour.

Variable volume terminal reheat systems supplied at ceiling level would also have a very large winter reheat load. If the terminal boxes turn the room air supply below six air changes per hour the buoyancy effect will stratify warm air at high level and give a totally unsatisfactory performance on heating. Such a system, operating in a 'free cooling' mode, would introduce very large quantities of fresh air, which would

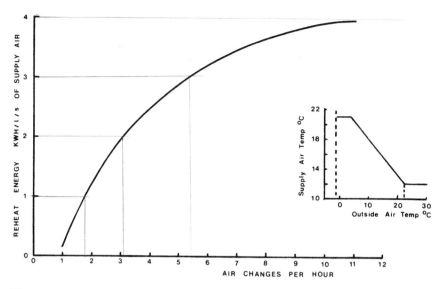

Figure 8.1   *Reheat energy during heating season to neutralize the cooling effect of fresh air in a free cooling system.*

be totally unnecessary in relation to the cooling requirements of the building and would absorb a large quantity of heating energy.

In an all-air system, or where the air conditioning system such as an induction system, requires three or four air changes of primary air, free cooling would require extensive reheat, but the basic net cooling requirement would be small. If we examine the basic cost of delivered energy, it can be seen that it saves energy cost to recycle the air and to recover the heat rather than admit extra fresh air. The cost per kWh of recycled energy from a refrigeration condenser is 1.42p/kWh which is considerably cheaper than the cost supplied by an oil-fired boiler, at 2.78p/kWh, or the cost supplied by a gas boiler, at 1.9p/kWh. The basis for the prices is as follows:

Electricity: 5.0p/kWh and COP of 3.5
Oil: 18.5p/litre and seasonal efficiency of 63%
Gas: 35.2p/therm and seasonal efficiency of 63%

The reheat energy in a free cooling system can best be illustrated by reference to Figure 8.1. The graph shows the energy consumption during the heating season to reheat the air to the room temperature per litre per second from a supply temperature indicated in the schedule, which is also indicated in Figure 8.1. The energy does not include the allowance necessary to heat the basic ventilation requirement.

Practical examples

When analysing a building with a view to installing an air to water heat pump, or, indeed, any air conditioning plant, it is first necessary to analyse the heating and cooling load in each room module. The variation in load throughout the year determines the characteristics of the distribution system. Having selected the distribution system, the central plant is considered, relative to the heating and cooling loads that would be imposed by the distribution system, throughout the year.

It should be emphasized that the cooling load imposed on the plant does not always reflect the actual cooling requirements of the building. It is often more a function of the distribution and terminal equipment control. For example, during winter operation of a variable air volume system with terminal reheat, the central air supply will, in all probability, be provided below room temperature and reheated. The temperature lift to room temperature is the unnecessary cooling load imposed by the system rather than the building.

The most convenient way of illustrating the effect that room loads have on the choice of distribution systems and, in turn, on the central plant, is by considering a number of practical examples. Four air conditioned buildings are presented which cover a wide range of design considerations. Metered heating and cooling energy is available and is used where appropriate to obtain design parameters, make comparisons and illustrate the various points.

Building No. 1 (Figure 8.2) [1] was completed in 1982 and is a two storey narrow plan brick building with cavity wall insulation and solid internal partitions. The building is cruciform in plan having four distinct wings. The main façades of the building face north-east, north-west, south-east and south-west and have double-glazed windows with an exterior pane of 6 mm 'Anti-sun' glass. The glazing is equally spaced in each face and is 29% of the total external wall area, when viewed from the interior. The air conditioned spaces are carpeted and have suspended ceilings with recessed air handling lighting fittings throughout. In the calculation of the cooling load the building has been considered as 'heavyweight' in connection with its thermal mass.

Building No. 2 (Figure 8.3) [2] was completed in 1975 and is a four

Figure 8.2   *Building 1: a two storey, well insulated low glazed area narrow plan office building completed in 1982.*

Figure 8.3   *Building 2: a four-storey low glazed area narrow plan office building completed in 1975 prior to part FF of the Building Regulations.*

storey narrow plan building constructed with precast concrete panels and having solid internal partitions. The building spine faces north-west and the two façades of the building face north-east and south-west. The windows are equally spaced, double glazed with 600 mm external window reveals. The glazing forms 30% of the total wall area in a typical building module. The air conditioned spaces are carpeted and have

Figure 8.4   *Building 3: a four storey speculative office building using 52% double glazing built in 1974.*

suspended ceilings with recessed air handling lighting fittings. In the calculation of the cooling loads the building has been considered as 'heavyweight' in connection with its thermal mass.

Building No. 3 (Figure 8.4) [3] was completed in 1974 and is a four storey narrow plan brick building, arranged generally with open plan offices. The plan form is 'L' shaped with two equal wings facing east and south. The four façades of the building face north, south, east and west and have equally spaced double glazed windows, with an area equal to 52% of the external wall, when viewed from the interior. The air conditioned spaces are finished with carpet tiles and suspended ceilings with surface-mounted lighting fittings. In the calculation of the cooling load the building has been considered as 'heavyweight' in connection with the thermal mass.

Building No. 4a (Figure 8.5) [4] completed in 1979, is a five storey narrow plan brick building arranged generally with open plan offices. The building is a simple rectangular block with a stairway and service core attached to the south-west elevation. The two glazed façades of the building face north-east and south-west with the main glazing area facing north-east. The windows are single glazed, are equally spaced and occupy 40% of the internal wall area. The air conditioned space has carpeted floors, and suspended ceilings with surface mounted lighting fittings. The building has been considered as thermally 'lightweight' for the purpose of calculating the cooling loads.

Figure 8.5   *Building 4: a five storey speculative office building using 40% single glazing completed in 1979.*

Building No. 4b is the same building as in Building 4a but the orientation has been changed so as to quantify the effect on the calculated cooling load. In the actual building 132 m² of glass faces north-east and 78 m² faces south-west. In this example the building has been reorientated so that the 132 m² of glass faces south-west and 78 m² faces north-east.

Physical details of the four buildings are summarized in Tables 8.1 and 8.2.

Table 8.1   *General details of buildings*

Building	Glazing area (%)	Thermal type	Lighting installed load (W/m²)	Gross area (m²)	Air conditioned area (m²)	Heated only (m²)	Ratio AC area cross area
1	29	Heavy	30	2853	2652	201	0.93
2	30	Heavy	18	5980	4837	1143	0.80
3	52	Heavy	12	3544	2992	552	0.84
4a & 4b	40	Light	13.5	1672	1417	255	0.85

Table 8.2   *Design heat loss*

Building	Calculated design heat loss (kW)	Design heat loss (W/m²) based upon gross area
1	106	37
2	320	54
3	215	61
4	92	55

The benefits of heat pumps are more apparent in buildings that conform to the current Building Regulations. This can be readily seen from the capital and energy cost given at the end of the paper. In the four examples Building 1 is within the Building Regulations and Building 2 generally agrees even though it was built 10 years ago. Buildings 3 and 4 are poorly insulated and have extensive glazing and are consequently outside the standards of the current Building Regulations. Nevertheless, they serve as good examples of the type of buildings that might be refurbished.

Building cooling load

In any particular building, for every zone with a high cooling load, there are other ones that require heating. Taking the building as a whole, there is a considerable diversity and overall the winter cooling load is small relative to the summer peak. To indicate the variation, the calculated seasonal cooling loads are given in Table 8.3 for the building examples. The loads take full account of diversity and make no allowances for the characteristics of particular systems.

A comparison of the calculated cooling load and the measured peak cooling requirements is given in Figure 8.6 for buildings 1 and 2. The excess capacity that is inherent in the calculation should discourage any tendency there may be to add a margin. On the contrary, it justifies the temperature swing allowance that reduced the cooling load by 10 W/m²; the space temperature swing being caused by the proportional band, or differential of the room control. It also indicates that the cooling load calculation should take full account of the thermal weight of the building.

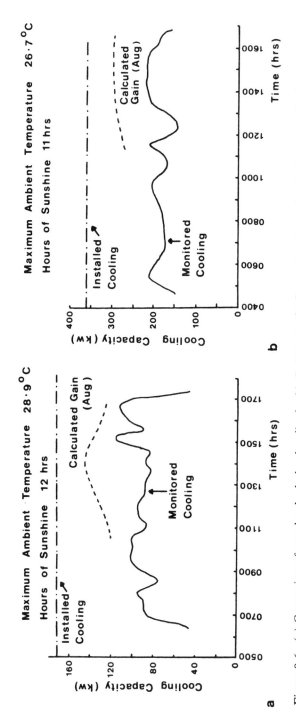

Figure 8.6 *(a) Comparison of actual and calculated cooling load in Building 1. (b) Comparison of actual and calculated cooling load in Building 2.*

Table 8.3   *Calculated building peak summer cooling load and peak loads in winter and intermediate seasons*

Building	Calculated cooling load (W/m^2) and percentage variation throughout the year					
	Summer		Intermediate (April)		Winter (March)	
	W/m^2	%	W/m^2	%	W/m^2	%
1	50	100	26	52	9	18
2	47	100	17	36	2	4
3	72	100	32	44	10	14
4a	84	100	38	45	19	23
4b	86	100	47	55	29	33

Table 8.3 indicates that the potential cooling load in April is of the order of 40–50% of the total plant capacity. In practice, the actual cooling load will be less than this, since the building cooling load calculations do not take into account the admittance of the building fabric. However, to ensure that sufficient cooling capacity is available in October and April, the heat pump plant would have to be arranged so that only a portion of the capacity changed over to heating and the remainder of the plant capacity remained in the cooling mode.

Regarding the winter cooling load, which, as indicated in Table 8.3, peaks in March, the required cooling capacity is negligible in buildings 1, 2 and 3. The cooling load calculations take into account the cooling effect of the normal ventilation requirements and an extra air change of fresh air at 12°C would absorb the cooling load of 10 W/m^2. It would be desirable in buildings 4a and 4b to have a small chiller available throughout the winter, although the utilization for cooling the building itself would be negligible. The main role would be heat recovery for the majority of the heating season with the occasional requirement for room cooling during March.

It is interesting to note, that whilst the calculations indicate that mechanical cooling is required during winter, the site measurements showed that no cooling was necessary or used from November to March in any of the four buildings. This is a further indication that all cooling load calculations should take account of the admittance of the building. This would certainly be a painstaking exercise, if carried out manually for the whole building, but is an obvious process for a computer.

An important point that is apparent when observing Table 8.3 is the magnitude of the summer cooling load in the highly glazed buildings 3 and 4 relative to buildings 1 and 2. The load is 50–60% greater. The same applies regarding the winter cooling load. The practical effect of this is high capital cost, system complexity and high energy cost.

It has long been established that excess glazing can lead to overheating in buildings. The corrective measures in terms of air conditioning are also expensive and the performance often unsatisfactory. It does not seem unreasonable to suggest that the area and type of glazing should be selected with due regard to the air conditioning system. As a general rule it can be claimed that the smaller the area of glazing, the simpler the air conditioning and the lower the capital cost and energy cost of heating and cooling.

Heat pump arrangements

All types of heating and cooling distribution systems can be engineered with heat pumps into an integrated system, but the capital cost would generally be minimized if the characteristics of the system and building were closely matched. Regarding the heat pump plant, this can be classified into two distinct arrangements, (a) and (b), with a hybrid system, arrangement (c).

(a) A changeover system with mechanical cooling in transitional seasons.
(b) A system providing heating and chilled water throughout the year.
(c) This system is similar to systems (a) and (b) but is more adaptable regarding the distribution systems to which it can be applied. It has a permanent heating and chilled water circuit together with a changeover circuit, which provides cooling in summer and on 1 October changes over to heating. This type of system would be suitable for buildings with extensive northerly zones. It would also be suitable for hotels, where the changeover system would be installed in the bedroom block and the four-pipe heating and cooling system would be installed in the public rooms. The heating circuit could also operate as a reheat circuit throughout the year, reheating the primary air. Providing reheat from a heat recovery chiller would be less expensive in energy cost than providing reheat from the heating plant as in system (a).

In the following sections the system hydraulics and controls are discussed in relation to the various central plant arrangements.

A general point regarding the hydraulic configurations for heat pumps is that they are arranged to provide primary and secondary circuits. There is a common cold feed for the whole system and all the pumps are situated to have their respective system pressure on the delivery side of the pump and a negligible suction pressure.

Primary and secondary pumping is desirable, so as to give direct mixing of water circuits, without one circuit having any hydraulic effect on another. Alternatively, heat exchangers could be used to separate the hydraulic circuits from each other, but they would require a temperature difference between primary and secondary which would have the undesirable effect in a heat pump system of downgrading the temperature.

Changeover system with mechanical cooling in transitional seasons

A system which would provide a suitable mode of control for this type of operation is indicated in Figure 8.7. The principle feature of this design is that at least two heat pumps are required. Referring to Figure

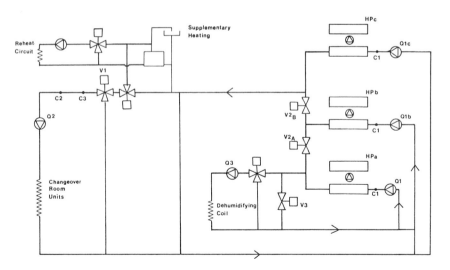

Figure 8.7   *Changeover system with mechanical cooling in transitional seasons.*

8.7, which indicates three heat pumps, all the heat pumps would operate on cooling during the summer, but on 1 October, at the changeover to winter operation, only heat pump HP_c would change over to heating. Heat pumps HP_a and HP_b would remain in the cooling mode. These two heat pumps would change over in sequence when the ambient temperature falls so that all three would be in the heating mode at 7–10°C, depending upon the characteristics of the building. The capacity of each heat pump would be selected by dividing the design summer cooling load by the number of heat pumps required. If three heat pumps were required, as in Figure 8.7, then the capacity of each heat pump would be one-third of the design cooling load. During the winter, from November to March, inclusive, the cooling load would be satisfied by fresh air, which is practically always below 10°C.

The hydraulic circuits are arranged and controlled so that the combined capacity of the heat pumps is available for each load. For example, if during the summer HP_a was out of commission, the cooling load imposed by the room units and the dehumidifying coil would be shared by heat pumps HP_b and HP_c. Referring to Figure 8.7, it may help to precede a description of operation by indicating the position of the two-position throughway valves V_2 and V_3 during the various times of the year. The temperatures indicated are typical and depend upon the characteristics of the building and the distribution system.

Valve	Summer	Intermediate Season >12°C	Winter Temp. Fall to 12°C	Winter Temp. Fall to 8°C
V_{2a}	Open	Open	Closed	Open
V_{2b}	Open	Closed	Open	Open
V_3	Closed	Open	Open	Closed

During the summer the heat pumps would operate on cooling. Each heat pump would be under the control of its own sensor C1, to produce chilled water at 7°C. The secondary circuit serving the room units would be controlled at 10°C, to provide sensible cooling only by means of sensor C2 and the modulating valve V_1. Pump Q_3 would serve the dehumidifying coil with chilled water at 7°C and control would be provided from the off coil temperature.

In order to prevent overcooling during the summer, the air off the dehumidification coil would be reheated under the influence of an ambient compensated temperature control, to a temperature not exceeding 28°C.

On 1 October heat pump HP_c would change over to heating and HP_b and HP_a would continue in the cooling mode, until the ambient temperature fell to 12°C. Below 12°C, the dehumidification and cooling requirement can be provided by fresh air and therefore the heat pump HP_b can be changed over to the heating mode. Value V_{2b} would open and valve V_{2a} would close. When the ambient temperature falls to 8°C heat pump HP_a would change over to the heating mode. Valve V_{2a} would open and valve V_3 would close. During winter operation pump Q_3 would not operate and the dehumidification control valve would remain permanently in the bypass position. If a number of dehumidification coils were installed, then each coil would have a separate control valve, rather than the one valve shown in Figure 8.7. In addition, an extra two-position throughway valve would be desirable on the dehumidification circuit, to prevent any water circulation during the winter.

Control of the heating during winter would be from sensor C_3, which would bring the heat pumps progressively on load and finally would control the supplementary heating valve. The three-way valve V_1 would be held in the straight-through position.

System providing heating and cooling throughout the year

Figure 8.8 indicates a four-pipe system which provides heating and chilled water throughout the year.

The heat pumps together with the heat recovery chiller would be selected so that the combined duty is equal to the calculated summer design cooling load. The heat recovery chiller would be selected to cater for the cooling load during the heating season from November to March inclusive. The hydraulic circuits are linked so that the capacity of each item of plant is available to supply the load imposed by each circuit.

The hydraulic configuration is similar to Figure 8.7 with the addition of the heat recovery chiller. The heat pump HP_c would change over from cooling to heating on 1 October and from heating to cooling on 1 May. The heat pumps HP_b and HP_a would change over in sequence from cooling to heating, as the ambient temperature falls as in system (a).

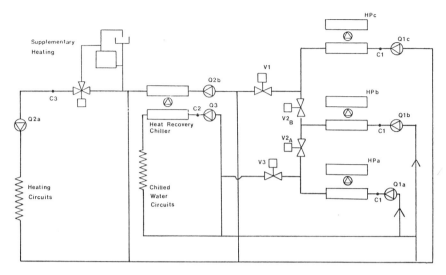

Figure 8.8   *System providing heating and chilled water throughout the year.*

The valves V_1 and V_3 are two-position valves and their positions at the various times of the year are given below.

Valve	Summer	Intermediate Season $>12°C$	Winter Temp. Fall to $12°C$	Winter Temp. Fall to $8°C$
V_1	Closed	Open	Open	Open
V_{2a}	Open	Open	Closed	Open
V_{2b}	Open	Closed	Open	Open
V_3	Open	Open	Open	Closed

During the summer, the heat pumps would operate in the cooling mode. Each heat pump and the heat recovery chiller would be under the control of its own sensor, C_1, to produce chilled water at 7°C. The chilled water would be supplied to the chilled water circuits under the power of pump Q_3. The return water from the chilled water circuits splits between the pumps Q_1a, Q_1b and Q_1c which serve the heat pumps. The heating circuits served by pump Q_2a are supplied from the heat recovery condenser and pump Q_2b. Valve V_1 would be closed. Valves V_2 and V_3 would be open.

On 1 October, heat pump HP_c would change over to heating, HP_a and HP_b would continue in the cooling mode until the ambient temperature fell below 12°C, at which point HP_b would change over to heating. In the intermediate period, when HP_a and HP_b are in the cooling mode, the cooling capacity available is the sum of HP_a, HP_b and the heat recovery chiller. Valves V_1 and V_3 would be open, V_{2a} would be open and V_{2b} closed.

When the ambient temperature falls below 12°C, heat pumps HP_b and HP_a change over to heating in sequence until all the heat pumps contribute to the heating load. Valves V_1 and V_2 would be open and V_3 closed.

Changeover system in conjunction with simultaneous heating and cooling

The system shown in Figure 8.9 would be applied to buildings that have substantial areas where a changeover system would be perfectly adequate, together with areas or circuits that require heating and cooling throughout the year. The hydraulic arrangement is similar to Figure 8.8 but with the additional changeover circuit.

The heat pumps together with the heat recovery chiller would also be selected as in arrangement (b), so that the combined duty is equal to the calculated summer design cooling load. The heat recovery chiller would

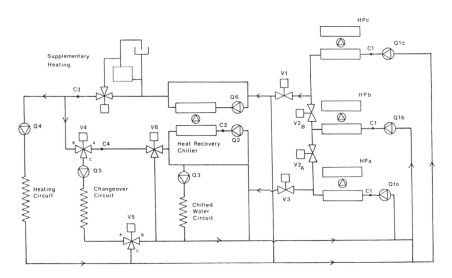

Figure 8.9   *Changeover system in conjunction with simultaneous heating and cooling.*

be selected to cater for the cooling load during the heating season from November to March inclusive. The hydraulic circuits are linked so that the capacity of each item of plant is available to supply the load imposed by each circuit.

The heat pump HP_c would changeover from cooling to heating on 1 October and from heating to cooling on 1 May in the same way as the system described in (b). Heat pumps HP_b and HP_a would changeover in sequence according to the fall in ambient temperature.

The valves V_1 to V_5 are all two-position valves and their positions during the various times of the year are given below.

Valve	Summer	Intermediate Season >12°C	Winter Temp. Fall to 12°C	Winter Temp. Fall to 8°C
V_1	Closed	Open	Open	Open
V_{2a}	Open	Open	Closed	Open
V_{2b}	Open	Closed	Open	Open
V_3	Open	Open	Open	Closed
V_4	AC	BC	BC	BC
V_5	AB	AC	AC	AC

During the summer the heat pumps would operate in the cooling mode. Each heat pump and the heat recovery chiller would be under the control of its own sensor to produce chilled water at 7°C. The changeover circuit, under the influence of pump Q_5, would be controlled at 10°C to provide sensible cooling only, by means of sensor C_4 and the modulating valve V_6. Pump Q_3 would serve the summer dehumidifying coils and any circuit that required chilled water throughout the year. Pump Q_4 would serve any summer reheat coils or other heating coils associated with those circuits providing simultaneous heating and cooling.

In the intermediate season and throughout the winter, heat pump HP_c is in the heating mode, commencing 1 October and terminating on 30 April. Heat pump HP_b and HP_a would change over to the heating mode in sequence as the ambient temperature falls below 12°C. The principle of the winter operation is that the first stage of heating in all heating circuits is provided by the heat recovery condenser. Subsequent stages of heating are provided by the heat pumps and finally by the supplementary heating plant. The heat pumps are loaded in sequence from the controller C_3 and on a fall in heating flow temperature due to insufficient heat pump capacity, the controller will provide supplementary heat, via the heating control valve.

Throughout the winter period the pumps Q_4 and Q_5 would serve the heating circuits and Q_3 would serve the chilled water circuits from the chiller. In a specific example, like Building No. 1, the loads given in Table 8.3 indicate that 52% capacity is required in April. Assuming that plant arrangement (a) is recommended for this project and that two machines are used, one machine would be in the cooling mode and one machine would be in the heating mode, at this time. Since the ambient temperature is no greater than 20°C, the cooling capacity available would be 15% greater than the design duty for the one machine. Relative to the peak summer capacity, the capacity available would be 57%, which is slightly in excess of the load.

Heating capacity

When considering the heating capacity required, it should be recognized that if the building is heated intermittently, a considerable overload capacity will be required. The overload capacity required for preheating, for the building examples, is indicated in Figure 8.10. When a short preheat period is given, the overload capacity is 50–75% of the design heat loss. Conversely, with a long preheat of 8 hours, the overload capacity required is no more than a nominal 10%. Heat pumps are of limited heating capacity and therefore, to make full use of the heat pump, the preheat period must be extended during cold weather. It is also desirable to make full use of the cheap-rate electricity available between midnight and 07.00 h.

Table 8.4 *Design day heating energy*

Building ref.	Design heat loss (W/m^2)	Preheat energy (kWh/m^2)	Daytime energy (kWh/m^2)	Total heating energy (kWh/m^2)	'Z'
1	37	0.33	0.30	0.63	0.70
2	54	0.21	0.70	0.91	0.70
3	61	0.20	0.86	1.06	0.72
4	55	0.23	0.71	0.94	0.72

The considerations for winter design conditions would be assisted by an analysis of metered heating energy consumptions. The energy consumption illustrated in Figure 8.10 is shown in Table 8.4. The item in Table 8.4 referred to as the 'Z' factor is the ratio of the energy used, relative to the design heat loss being maintained at the same output, for

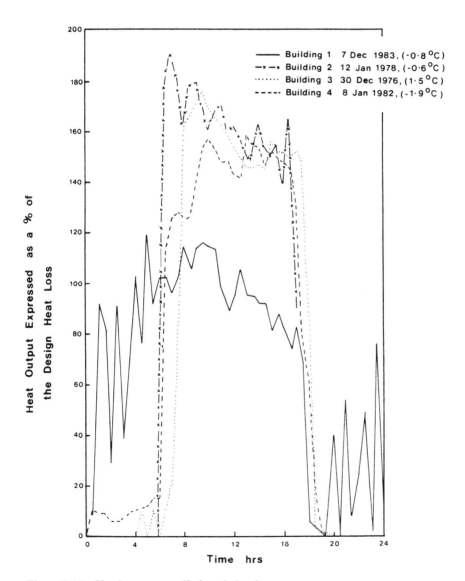

Figure 8.10 *Heating output profile for a design day.*

the full 24 hours. Since the 'Z' factors are similar, it can be concluded that the energy consumption for a design day is approximately equal to 17 hours full load operation, at design heat loss. Assuming the preheat commences at 01.00 h, it is reasonable to allow 6H to be supplied to the building before 07.00 h, when the low-rate electricity charge period ends.

The energy input can be summarized as follows:

H = design heat loss at −1°C kW
HP = heat output of heat pumps at −1°C kW

Assume the supplementary heat is provided by electric thermal storage and the storage water temperature drop is 60 K.

Total heat energy   = 17H kWh

Preheat energy
prior to 07.00 hours   = 6H kWh

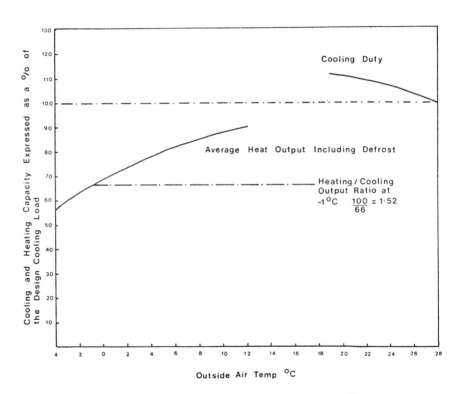

Figure 8.11 *Typical cooling and heating output curves of air to water heat pumps.*

Supplementary heat
energy during preheat   = 6(H-HP) kWh

Supplementary heat
energy during day   = 11(H-HP)

Storage in litres $= \dfrac{11(\text{H-HP}) \times 3600 \text{ litres}}{4.2 \times 60 \text{ K}}$
of water

= 157(H-HP) litres

Electricity charge rate   = 2.43(H-HP) kW
at night

The rated output of the heat pump at $-1°C$, in the heating mode, should take into account the energy lost in defrost and also the time lost in heating while the heat pump is in the defrost mode of operation. The net heating output can be expressed relative to the output in the cooling mode during summer design conditions.

Data from field trials given in Figure 8.11 indicates that the output on cooling expressed in kW of cooling is 1.52 net output on heating at $-1°C$.

Cooling output = 1.52 (heating output) kW.

Distribution system considerations

In order to illustrate how heat pumps can be applied universally in air conditioning it is necessary to describe how they can be integrated into a complete system in the various types of building.

They should also be related to the more common central plant distribution systems in general use at the present time. These are as follows:

1 Two-pipe perimeter fan coil system, supplying sensible cooling and heating, with separate high level ducted air supply, providing dehumidification and ventilation. Alternatively, the fan coil units could be mounted at high level in the corridor void.
2 Two-pipe changeover induction system.
3 Variable air volume air supply at high level, with terminal reheat.
4 Variable air volume air supply at high level, with perimeter radiator or convector system.
5 Four-pipe induction system.
6 Four-pipe fan coil system.

The four-pipe systems, items (5) and (6) can be universally applied, but in many buildings would be under-utilized. There is, therefore, a risk of an unnecessary high capital expenditure. The two-pipe fan coil system, item (1), is undoubtedly the lowest in capital cost, but for its successful application, a more accurate prediction of the seasonal heating and cooling loads is necessary. The two-pipe changeover induction system, item (2), is a more expensive variation of the two-pipe fan coil system.

The variable air volume system with terminal reheat, item (3), has practical limitations, regarding the duct sizes that may be necessary in highly glazed buildings. In addition, there is a limit to the turn down that can be permitted when in the heating mode. Usually a minimum of six air changes per hour is required to provide heating successfully at high level. This would necessitate a large reheat load at all times, which would give rise to high energy costs unless heat recovery was employed. The variable air volume system with a perimeter radiator system, item (4), is a better choice relative to the performance and energy consumption. The VAV system could turn down to two air changes and the low level heating system would certainly provide a more satisfactory heating performance.

To illustrate how the relevant distribution system and plant arrangement is selected, each of the four building examples is analysed in turn.

Building 1

For 29% glazing, in a south-west module, the intermediate season cooling load for that zone is 22 W/m² and 10 W/m² in March. These calculated cooling loads take into account the admittance of the structure. Assuming first of all, a two-pipe fan coil system, which was the actual system installed in the building, the first consideration is the air supply during the intermediate season. Since the fan coil water distribution is in the heating mode the cooling requirement is provided by the primary air. If plant arrangement (a) is adopted (see Figure 8.7) and assume that two heat pumps are used; one heat pump would operate on heating from 1 October until 30 April and the second heat pump would operate on heating at ambient temperatures below 10°C. If the air is supplied to each module at 12°C and at peak April load the temperature rise of the supply air is 12 K, the air supply quantity should be 1.0

litre/m²s, which is less than two air changes per hour. In March the building heat gain coincides with an ambient temperature of 12°C maximum and the heat gain can be satisfied by the two air changes of fresh air. Therefore, throughout the heating season, two air changes of ducted air should be controlled at a supply temperature to the room of 12°C. When the ambient temperature falls below 10°C the fresh air can provide all the necessary cooling to the building.

If a changeover induction system were preferred, the design considerations would be exactly as above. A variable air volume system with terminal reheat would need to be associated with plant arrangement (b). A minimum of six air changes would be supplied at all times to provide heating and the cooling and reheat would be provided by the heat recovery chiller. The duty of the chiller would be just sufficient to cater for the winter load and would be capable of cooling six air changes by 2.5 K. The system would be provided with a fixed minimum fresh air quantity throughout the heating season. It would be preferable to have a ceiling-mounted variable air volume system in conjunction with a perimeter heating system, which would also operate with plant arrangement (b). Since the air system can turn down to two or three air changes the system is in effect similar to a changeover fan coil or induction system.

The four-pipe induction or fan coil systems would also operate similar to a two-pipe system. In a building of this type, the two chilled water pipes would be redundant during the heating season. The primary air would have sufficient cooling capacity to provide the necessary cooling.

Building 2

The considerations concerning the choice of system are the same as for Building 1. The system actually installed in this building is a two-pipe changeover induction system.

Building 3

In reality a unitary air cooled system is installed in this building. At the design stage, it was considered that the large area of south facing glass would require cooling throughout the year. In practice, cooling was not required during the period November to March. Referring to the south facing module for 52% glass, the calculated cooling load for April is 66 W/m² and 65 W/m² in March. Loads of this magnitude rule out the simple changeover systems. It would be necessary to make a choice between systems (3), (4), (5) and (6). On the north and east zones the system could be simplified. If a four-pipe fan coil or induction system were installed on the south and west zones a two-pipe system would be adequate on the north and east zones. It is unusual to mix systems in this manner but the performance would be satisfactory and it would save capital.

Regarding the central plant, two heat pumps in conjunction with a heat recovery chiller would provide the necessary capacity during the heating season. Arrangement (b) (see Figure 8.8) would be a suitable plant arrangement unless the two-pipe system is considered on the north and east zones, in which case arrangement (c) (see Figure 8.9) should be used.

Buildings 4a and 4b

Because of the south-west orientation and the large area of glazing in each module the design considerations are the same as for Building 3.

System choice

There are other reasons to be taken into account apart from the thermal considerations when making a choice of air conditioning system. There are the obvious ones of capital cost, energy cost, maintenance and aesthetics; and there is also the added consideration of floor space. Perimeter systems are often excluded because they occupy useful lettable floor area. The object of this paper is not to indicate any preferences but to show that all systems can be engineered into heat pump installations and give capital and energy cost comparisons. However, it is first necessary to indicate the plant requirements of the various systems.

Table 8.5 indicates the heat pump plant arrangements that are necessary with the various distribution systems together with the plant duties. Table 8.5 shows that the distribution systems do not affect the installed cooling capacity but that they do affect the type of plant.

Table 8.5 *Heat pump plant arrangements for the various buildings and systems together with plant duties*

System type	Heat pump arrangement	Number of heat pumps	Installed cooling capacity (kW)	Heat recovery machine duty (kW)	Heat output of heat pumps (kW)	Thermal storage capacity (litres)
Building 1						
i	a	2	142	NR	93	2,300
ii	a	2	142	NR	93	2,300
iii	b	2	142	36	70	5,900
iv	a	2	142	NR	93	2,300
v	a	2	142	NR	93	2,300
Building 2						
i	a	2	281	NR	185	21,000
ii	a	2	281	NR	185	21,000
iii	b	2	281	76	135	29,000
iv	a	2	281	NR	185	21,000
v	a	2	281	NR	185	21,000
Building 3						
iii	b	2	215	40	115	16,200
iv	b	2	215	27	124	14,900
v	b	2	215	27	124	14,900
v Mod	c	2	215	27	124	14,900
Building 4a						
iii	b	2	119	44	49	6,900
iv	b	2	119	21	64	4,500
v	b	2	119	21	64	4,500
v Mod	c	2	119	21	64	4,500
Building 4b						
iii	b	2	122	44	51	6,600
iv	b	2	122	35	57	5,600
v	b	2	122	35	57	5,600
v Mod	c	2	122	35	57	5,600

Systems requiring reheat require a heat recovery machine for winter operation even though the building has no significant winter cooling load. This applies to Buildings 1 and 2. Buildings 3 and 4, however, have a high winter cooling load because of the extensive south facing glass. The variable volume distribution with terminal reheat also requires a significant increase in the supplementary heating requirements. This is due to the incorporating of a heat recovery chiller and the reduction in heat pump capacity.

Capital cost

There is a big variation in capital costs between the systems indicated in Table 8.5. Systems (3), (4), (5) and (6) are more complex than systems (1) and (2) and are correspondingly more expensive. The highly glazed Buildings 3 and 4 require the more complex systems and as can be seen, systems (1) and (2) have been excluded in these two buildings. Buildings 3 and 4 also require large plant capacities for a given building area which also increases the capital cost.

The comparison with conventional refrigeration plant and boiler

Table 8.6   *Capital cost of heat pump systems ($£/m^2$)*

| | System type and cost ($£/m^2$) | | | | | | |
ia	ib	ii	iii	iv	v	vi	vi mod
Building 1							
78	97	111	128	135	132	115	N/A
Building 2							
77	96	108	126	132	128	114	N/A
Building 3							
N/A	N/A	N/A	148	156	152	128	119
Building 4a							
N/A	N/A	N/A	135	140	146	126	114
Building 4b							
N/A	N/A	N/A	135	141	147	127	115

Key
ia   two-pipe fan coil with high level corridor distribution and terminals
ib   two-pipe fan coil with perimeter terminals
ii   two-pipe changeover induction system
iii   variable air volume at high level with terminal reheat
iv   variable air volume at high level with perimeter radiator or convector system
v   four-pipe induction system
vi   four-pipe fan coil system
vi mod mixture of four-pipe and two-pipe fan coil system

Prices are based upon the gross area of the building.

plant is much more difficult. The differences are so small that they cannot be identified without an inordinate amount of research. Consequently the comparisons have been ignored.

Table 8.6 gives the capital cost of the heating and air conditioning systems indicated in Table 8.5. The unit costs are based upon the gross building areas given in Table 8.1.

The greatest potential for reducing capital is by designing a building with reduced glazing, as in Buildings 1 and 2, in which the simple fan coil system, 1a, can be installed. By comparison, system 4 in Building 3 is twice as expensive. When the more complex systems are applied in Building 1, the costs escalate but the attributes of the systems are superfluous to the requirements of the building.

Building energy cost

The energy cost of the heat pump systems is compared with systems using conventional refrigeration plant and boiler plant. Where heat recovery plant is incorporated into the heat pump system it is also incorporated into the refrigeration machines used in the comparative systems. The energy consumption and costs are given in Tables 8.7 and 8.8. In order to give a realistic comparison the total energy consumption and cost for the whole building is given. This includes lighting, power, lifts and all services. The energy consumption for each building is based upon the metered data for the services and space heating. The metered data serves as a base upon which adjustments are made in energy consumption and electricity maximum demand according to the characteristics of the system.

The tables should not be used to compare the energy cost of one building against another because the lighting loads and other loads are unequal. The tables should be used to compare the systems and fuels in the same building. The building incorporating the heat pump is approximately $5p/m^2$ to $10p/m^2$ lower in energy cost than when heated by gas and between 70p and £1.00 per m^2 lower than when heated by oil.

The type of air conditioning system used also influences the energy cost. The energy cost of system (3), the high level variable air volume terminal reheat system, is approximately 60p to $77p/m^2$ higher than the next highest energy cost. In Buildings 1 and 2 it is £1.00/m^2 higher than the system bearing the lowest energy cost.

Table 8.7 *Building energy consumptions comparisons*

System	Heat pump systems				Gas or oil heated system			
	Electricity				Fuel			
	kWh/m^2		W/m^2		kWh/m^2		W/m^2	
	Night	Day	Night	Day	Night	Day	Night	Day
Building 1								
i	27	94	32	46	9	74	32	120
ii	27	97	32	48	9	77	34	120
iii	42	112	55	55	13	95	44	114
iv	26	96	32	46	8	76	32	120
v	30	99	32	49	11	79	35	123
vi	27	96	32	48	9	76	34	120
Building 2								
i	55	74	75	38	13	55	26	149
ii	55	77	75	39	13	58	27	149
iii	69	90	98	46	17	76	37	143
iv	54	76	75	38	12	57	26	149
v	58	79	75	40	15	60	28	157
vi	55	76	75	39	13	57	27	149
Building 3								
iii	55	80	97	48	9	62	34	153
iv	47	64	83	45	4	44	26	161
v	51	68	83	45	7	47	30	177
vi	49	64	83	42	6	44	27	161
Building 4a								
iii	58	104	89	51	13	86	38	150
iv	41	82	62	45	7	59	28	158
v	46	92	62	49	11	68	32	173
vi	43	90	62	47	9	67	30	158
Building 4b								
iii	58	105	87	51	13	87	38	150
iv	47	81	70	43	7	60	28	158
v	52	91	70	47	11	69	32	173
vi	49	89	70	45	9	68	30	158

Day demand Profile %	Jan.	Feb.	Mar.	Apr.	May.	Jun.	Jul.	Aug.	Sep.	Oct.	Nov.	Dec.
	100	100	65	65	70	75	75	75	75	75	80	100

Night demand Profile %	Jan.	Feb.	Mar.	Apr.	May.	Jun.	Jul.	Aug.	Sep.	Oct.	Nov.	Dec.
	100	75	10	10	10	10	10	10	10	50	75	100

Table 8.8   *Building energy cost comparisons (£/m²)*

System	Total building energy cost (£/m²)		
	Heat pump system	*Gas heated*	*Oil heated*
Building 1			
i	5.21	5.28	5.94
ii	5.36	5.44	6.10
iii	6.31	6.30	6.93
iv	5.26	5.33	5.99
v	5.51	5.62	6.29
vi	5.32	5.40	6.07
Building 2			
i	4.81	4.85	5.72
ii	4.94	4.99	5.86
iii	5.81	5.84	6.67
iv	4.86	4.90	5.77
v	5.08	5.21	6.13
vi	4.91	4.95	5.82
Building 3			
iii	5.31	5.34	6.19
iv	4.44	4.54	5.43
v	4.74	4.99	5.97
vi	4.50	4.60	5.49
Building 4a			
iii	6.50	6.54	7.37
iv	5.28	5.35	6.23
v	5.82	6.00	6.96
vi	5.64	5.72	6.60
Building 4b			
iii	6.54	6.58	7.41
iv	5.31	5.39	6.27
v	5.83	6.04	7.00
vi	5.68	5.75	6.63

Gas oil 18.5p/litre
Gas 35.2p/therm

Heat pump system energy cost

Table 8.9 indicates the cost in p/kWh of the heat delivered in the various heat pump systems. The unit cost is the average cost taking into account the input from the heat pump and the supplementary heat supplied by the thermal storage.

Table 8.9   *Average cost of heat delivered in heat pump system*

System	Heating energy output (kWh/m²)	Percentage of total heat delivered (%)			Thermal storage & heat pump system contribution to MD (W/m²)		Average cost of heat delivered
		Thermal storage	Heat pump		Night	Day	(p/kWh)
			Night	Day			
		All night	input × $\frac{1}{2.22}$	input × $\frac{1}{2.44}$			
Building 1							
i	80	7.2	10.8	20	26	14	1.69 ⎫
ii	80	7.2	10.8	20	26	14	1.69 ⎪
iii	76	18.2	10.3	15.9	42	11	1.76 ⎬ 1.70
iv	80	7.2	10.8	20		14	1.69 ⎪
v	82	7.4	11.0	20.5	26	14	1.68 ⎪
vi	80	7.2	10.8	20	26	14	1.69 ⎭
Building 2							
i	99	32	22	46	69	12	1.70 ⎫
ii	99	32	22	46	69	12	1.70 ⎪
iii	95	46	18	36	85	9	1.72 ⎬ 1.70
iv	99	32	22	46	69	12	1.70 ⎪
v	101	32	23	45	69	12	1.69 ⎪
vi	99	32	22	46	69	12	1.70 ⎭
Building 3							
iii	102	35	22	43	84	14	1.75 ⎫
iv	107	30	23	47	79	15	1.74 ⎬ 1.74
v	110	29	24	47	79	15	1.74 ⎪
vi	107	30	23	47	79	15	1.74 ⎭
Building 4a							
iii	100	35	21	44	76	13	1.76 ⎫
iv	105	21	26	53	58	17	1.75 ⎬ 1.75
v	107	21	26	53	58	17	1.74 ⎪
vi	105	21	26	53	58	19	1.75 ⎭
Building 4b							
iii	100	35	21	44	74	13	1.76 ⎫
iv	105	28	24	48	66	15	1.74 ⎬ 1.74
v	107	27	24	49	66	15	1.73 ⎪
vi	105	28	24	48	66	15	1.74 ⎭

In general terms the greater the area of glazing the higher the cost of energy delivered. The reason for this is that a chiller is required during the winter months to satisfy the systems that are applied in highly glazed buildings. Consequently there is less heat pump capacity available for heating. This can be seen in Buildings 3 and 4. The average cost of heat delivered by the heat pump system in conjunction with the supplementary heat provided by thermal storage is as follows:

Building 1	1.70p/kWh
Building 2	1.70p/kWh
Building 3	1.74p/kWh
Building 4a	1.75p/kWh

By comparison, the cost of useful heat in the occupied space when generated by an oil or gas boiler is 2.78p/kWh by oil and 1.9p/kWh by gas.

Conclusions

The air to water heat pump can be engineered into any type of air conditioning system and applied to buildings of all types. The capital cost advantage against more conventional refrigeration equipment in association with oil- or gas-fired boiler plant has not been investigated in detail but is considered to be negligible.

The capital cost benefit does not lie between the choice of heat pump and conventional refrigeration and boiler plant but in the choice of system together with the thermal characteristics of the building. There is a factor of two in the range of capital costs of the air conditioning when considering the different buildings and systems; £77–78/m² is the lowest capital cost in Buildings 1 and 2 and £156/m² the highest in Building 3.

The difference in energy cost between systems is as great as the difference in energy cost across the three fuels; approximately 70p/m² to £1.20/m². The highest energy cost system using oil is at least £1.70 to £2.10/m² greater than the lowest energy cost system using a heat pump for the same building.

When considering the capital costs and the energy cost comparisons it can be concluded that effort should be directed towards designing buildings with thermal characteristics similar to Buildings 1 and 2 and reaping the capital cost benefit. Simple changeover air conditioning systems can then be installed with capital costs no greater than £78/m². Heat pumps are an ideal choice in changeover systems since they can be utilized throughout the year and also obtain the minimum energy cost.

References

1   Adams D. F., Leary J. & Austin B. S. (1984) The application of central air to water heat pumps in a traditionally constructed well-insulated narrow plan building. *Heat Pumps in Buildings* (Hutchinson).

2   *Energy consumption and thermal performance of a well-insulated narrow plan office building with a central air conditioning system*. Environmental Engineering Section. The Electricity Council. ECR/R1468.

3   *The recorded energy consumptions and performance of an all electric unitary air conditioning system in an office building*. Environmental Engineering Section. The Electricity Council ECR/R1125.

4   *All electric unitary air conditioning system in an office building in Romford*. Environmental Engineering Section. The Electricity Council.

Discussion

M. D. Terry (North Thames Gas)   Combined heat and power in conjunction with double-effect absorption chillers would on the surface appear to have merit particularly in running costs, although capital costs are likely to be a little higher. Such a system could satisfy both heating, chilling loads and power generation requirements.

B. A. Bath (Rybka Smith & Ginsler, Toronto)   We have used absorption chillers in hospital applications over the years and have found the life expectancy of an absorption chiller to be only about 10 years, so the cost, based on this short life expectancy, is a very serious consideration.

J. Leary (Electrical Council)   The balance of energy is still a problem. Absorption chillers are heavy users of energy. For every 1 kW of refrigeration some 1.5 kW of fuel is needed. If the boiler efficiency is taken into account some 2.0–2.2 kW is needed to produce 1 kW of refrigeration. To combine an absorption machine with an electricity generation set would be a difficult task with zero return on capital. Reliability is also a problem. Absorption machines are complex and adding local generation from a diesel or gas engine just adds to the complexity. It also seems unlikely that the electricity could be produced cheaper than the cost of purchasing it from the National Grid. Only part of the heat will be used from the generator and the remaining heat would be surplus and rejected. In Chapter 8 we show how little energy well-designed, modern buildings need – they could not possibly be matched by high-energy intensive pieces of equipment such as absorption machines and electric generation sets.

E. J. Perry (W. S. Atkins & Partners)   From computer models of vapour compression systems and absorption systems using waste heat, the vapour compression system would appear to be far more economical than the simple absorption system.

Dr J. Paul (Sabroe & Co.)   The heat balance is very much related to duties in the particular project. A successful example was a gas-driven heat pump installed in a Munich ice rink. There was surplus heat from the drive at times of no heating demand, so the design combined one reciprocating compressor and one screw compressor, both driven by gas engines, with one screw and two reciprocating compressors with electric drive. For freezing the ice rink in late summer all the machinery is operated and there may be a couple of days when there is surplus heat which is removed by a cooling tower. In the autumn and spring seasons with low heating demand the electrically-driven equipment is used, but in winter with high heat demand the installation goes over to the gas drive.

Less successful was an installation in a leisure centre with a swimming pool and an ice rink. It was quite clear that by using a gas drive the pool could be heated and the ice rink frozen at the same time. After 3 years of operation the gas engine has been abandoned in favour of electic motors, because the waste heat from the engine had in practice to be rejected using a cooling tower.

There is an alternative way of using waste heat: applying organic Rankine cycles. We are running a total of three units which were designed without external funding and are operated purely on economic grounds. Waste heat from incineration plants is at a temperature of 150–160°C which is too high for use directly with a heat pump. This heat is used to drive an organic Rankine cycle engine coupled to an electricity generator and the exhaust heat is fed into a district heating network. Three plants have been commissioned so far and they are all economic. The capital investment per kW electric output is roughly the same as that for a normal power station and in one case lower. These developments are only just starting – costs per kW output will go well below costs in a steam power station even though organic cycles have an efficiency of only 9–12%.

D. C. Pinto (Temperature Ltd)   Chapter 8 gives figures for a series of five buildings with systems involving central station heat pumps. Have comparative figures been obtained from unitary heat pumps in similar buildings?

J. Leary (Electricity Council)   Unitary heat pumps were examined but because energy economy is the aim, only unitary heat pumps using a heat source exterior to the building were considered. The particular unit manufactured by Mr Pinto's company, the 'Versatemp', was not therefore included, as in that case the energy has to be supplied from a boiler. Comparative energy cost figures including 'Versatemp' are available for buildings of this kind, and of all the very many systems analysed the variation is of the order of £1/m². It does not seem likely that the 'Versatemp' could be cheaper because of the way the boiler output energy has to be supplied to the building via the heat pump.

We would like to see an external source unitary heat pump: a through-the-wall unit – just an external condenser and supplementary electric heating. The energy costs of such a unit would be similar to the quite reasonable costs shown in Table 8.4, and the capital cost would be much reduced. An external source unitary heat pump system would be cheaper than a central station system, but there is a problem of handling defrost water to prevent it running down the outside of the building.

E. J. Perry (W. S. Atkins & Partners)   The cost per kilowatt hour of recycled energy from a refrigeration condenser is given as 1.42p/kWh with electricity at 5p/kWh and COP of 3.5. Is 1.42 referred to as the heat pump coefficient performance?

B. S. Austin (Electricity Council)   The figure of 1.42p/kWh is the actual cost of heat supplied by the heat pump. It is derived by dividing the cost per kWh of electricity by the COP.

9 New UK air conditioned buildings using heat pumps

K. J. V. Fowler

Summary

This chapter outlines the trends of building and services costs and technology over the past two decades which have encompassed the years 1973 and 1974 when the price of oil increased by some 600%. The increase continued through the second half of the decade and, understandably, produced a consumer demand for other, less costly fuels.

The immediately viable alternative was gas but, because North Sea production was not fully developed, supplies were limited. This factor, coupled to the severe winter of 1978/79, obliged the British Gas Corporation, in the autumn of 1979, to impose restrictions on supplies to new developments: the restrictions limited new connections to a site boundary distance of 25 m from the low-pressure gas main and to a maximum annual consumption of 25,000 therm.

Many building developments under way at the time were unable to meet these requirements and it was then that the heat pump began to receive wider consideration as a solution to the supply crisis and, with other forms of energy conservation, as a means of curbing the rapidly increasing running costs of building services.

The principle of the heat pump has been well explained elsewhere but its application remains to be fully exploited. Hence, this chapter describes several recent air conditioning installations using heat pumps in various forms together with other components used for energy conservation.

Introduction

The post-war building expansion accelerated some 20 years ago under relatively stable economic conditions: the retail price index had hovered around 3.5% since the war; the male average wage was £20/week, and oil was plentifully available at 1 shilling per gallon (1p/litre in present-day terms).

Related to net lettable areas, average office rents in the West End of London (where the expansion began) were £26.90/m² for 99 year leases with reviews after 21 years; the installed costs of the mechanical and electrical services in air conditioned buildings were of the order of £28/m² and the annual running costs averaged £1.27/m² – the equivalent of 4.7% of the rental values.

Corresponding figures today are £194/m² for rental values, £226/m² for mechanical and electrical services installed, and £7.20/m² for the associated running costs. Hence, whilst the relationship between installed costs and rental values has shown a slight increase over the period, running costs at 3.7% of rental values are proportionately lower than they were in 1964.

This running-cost reduction of 1%, which has been achieved in the face of the dramatic increases in energy costs that have occurred over the period, is a result of the improved standards of building construction and insulation that have been introduced since 1973, and of the new technology and equipment which is now commonly available to the building services industry.

Specimen calculations for buildings recently completed show that, had these improvements not been sought and applied, the running costs of the installed services would have escalated to an annual average rate of £11.80/m² which is the equivalent of 6.1% of the average rental values.

Whilst London rents in themselves are not among the world's highest, they become so when expressed as elements of total occupation costs, that is, with the additions of rates and service charges. Comparative occupation costs for the City of London, New York, Tokyo, and the central areas of London's West End are, respectively, £567/m², £491, £479 and £374.

Analysed, the figures for the City and central West End appear something like this:

	City	West End
Rents	£333.5/m²	£236.7/m²
Rates	£191.5/m²	£113.5/m²
Service charges	£ 42.0/m²	£ 23.7/m²
Occupation costs	£567.0/m²	£373.9/m²

As percentages of the rental values the rates and service charges are:

	City	West End
Rates	57.5%	48%
Service charges	12.5%	10%
Totals	70%	58%

Although latterly, rents have stabilized to some extent, the on-costs are progressively increasing. This is a cause of concern for the whole of the development team, but particularly for the building services engineer: there is nothing that he can do to control the ever increasing level of rates but, within the limits of his jurisdiction, he can, and is, keeping in check the service charge costs.

In this latter respect the priorities for the services engineer are:

- to exert influence for the building design not only to be thermally efficient but also to be as airtight as possible; leaking window frames can impose a severe penalty on winter running costs.
- to minimize running costs by the use of energy-saving designs, systems and equipment.
- to limit service, depreciation and sinking-fund costs by the selection of equipment which is robust, durable and easily serviced.

Since 1964, and particularly following the energy crises of the 1970s, significant advances have been made in building services technology and the equipment available to designers:

- computer-aided designs are more efficient and cost-effective.
- new and more effective systems such as the multi-pipe, variable-air-volume and Versatemp systems are in common use.
- among the new equipment are heat pumps, heat pipes, thermal wheels and other economizers; high-efficiency lighting systems; energy management units; selenium-cell controllers, and so on.

Each of these improvements has contributed to the containment of first costs and service charges that has been achieved, but the greatest assistance in these respects has resulted from the availability of purpose-made heat pumps.

Heat pumps have been used periodically over the past 50 years with air, water, grain, soil and sewer sources of low-grade heat. They were of an experimental nature using standard refrigeration assemblies adapted for reverse-cycle operation; the air source versions requiring auxiliary heating at outdoor temperatures of 5°C and below.

Air and water being the most readily available and effective heat sources, packaged assemblies using these media are now marketed by many UK and overseas manufacturers, and installations using them have been in service for a number of years, generally independently of auxiliary heating or, in some instances, using available heating for standby purposes.

Heat pumps are viable in use with most distribution systems (induction, fan coil, Versatemp, etc.) as the following installation examples will show.

Versatemp system with central air to water heat pumps

Figure 9.1 is a photograph of a building for which the instructions were to design the mechanical and electrical services and suspended ceilings; the design to include Versatemp air to water heat pump units.

In 1979 the high demands for gas as an alternative to oil obliged British Gas to invoke their statutory supply conditions with which the new building could not comply and the overall costs of using oil were too high for the applied budget.

Figure 9.1   *Building using Versatemp system with central air to water heat pumps.*

These factors, together with the specification for using the Versa-temp system, pointed the design towards air source heat pumps in place of conventional boilers. From the outset of development of the Versa-temp idea it was apparent that such a combination would result in a lower temperature of the primary hot water circuit, more in line with the level of 24°C required for the Versatemp system; would introduce heat reclaim in addition to heat transfer; and would result in a more useful coefficient of performance.

Figure 9.2 is a diagrammatic representation of the Versatemp system.

The system comprises individual self-contained air conditioners incorporating an hermetically-sealed refrigeration assembly, a fan and motor, two heat exchangers 'A' and 'B', an air filter and in-built or remote controls.

Water at a constant nominal temperature (24°C) and constant flow rate is circulated through heat exchanger 'B' in each unit.

For cooling purposes the refrigerant reversing valve is in position 1. Heat removed from the room air at heat exchanger 'A' is rejected into the water circuit at heat exchanger 'B'.

For heating purposes the reversing valve is in position 2 and the unit is now acting as a 'heat pump'. Heat is extracted from the water at heat exchanger 'B' and is transferred through the refrigeration circuit to the room air via heat exchanger 'A'.

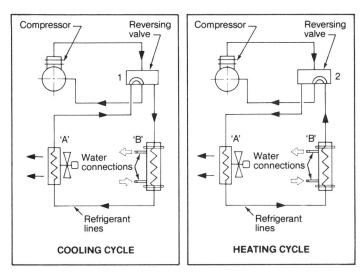

Figure 9.2 *Diagrammatic cycles of Versatemp cooling and heating circuits.*

The changeover from cooling to heating is effected automatically by the unit controls.

A detailed analysis was made of the running costs of the system compared with oil and gas firing; the analysis, in each case, being comprehensive of all services and including secondary forms of heat recovery from lighting, waste air, etc. Before placing the results before the client, the analysis was given to the Electricity Council for checking and comment.

The results of these analyses showed that the use of heat pumps in place of conventional boilers would produce running costs below those for either gas or oil heating. For various reasons the Electricity Council figures differed from our own, but the comparative increases in the running costs of oil and gas firing compared with heat pumps were relatively similar, thus:

	K. F. & P.	*Electricity*
		Council
Gas-fired system	+ 31%	+ 27%
Oil-fired system	+ 75%	+ 71%

Provisional budget costs indicated that the installed costs would also be lower than for systems using conventional plant.

The design comprises a basic Versatemp system of 240 units working in series with four 64 kW air to water heat pumps, together with other refinements for energy conservation, thus:

- a heat wheel in the air plant, speed controlled by a supply air thermostat.
- an energy management unit which monitors all elements of the electrical demand and which is programmed to switch out non-essential circuits of the demand load as the maximum is approached.
- the lighting forms part of the load shedding. Hence, a dual-circuit arrangement is used each serving high-efficiency fluorescent tubes mounted one above the other in the fittings; the upper tubes being connected to the load-shed circuit. This limits shadow effect in the fittings so that the reduction of illumination is scarcely noticeable.
- the fittings are pre-wired to render them self-contained and inter-changeable with each other or with spacer fittings and are used to carry the ceiling thus eliminating the normal form of suspension.

The EMU also controls the following functions:

Plant running   The Versatemp units are phased in and out in groups through the building in order to limit the impact loads on both the electrical system and the central heat pumps and to match the heating and cooling input with the prevailing building demand.

Night set-back   With the lighting switched off, the plant is programmed to hold a winter night set-back temperature of 10°C. At this setting the ventilation plant runs on fully recirculated air, the heat-wheel is switched off and the Versatemp system is phased in only when required.

Boost start-up   With the ventilation plant on full recirculation, a boost heater is brought in at a time prior to occupation sufficient to allow the design temperature to be reached according to the prevailing outdoor temperature. At occupation, the plant is switched to normal running condition.

Standby generator   On mains failure the generator starts automatically and the waste heat is transferred to the Versatemp water circuit for useful heat recovery, or rejected to the cooling tower according to season.

Other aspects of the design are:

Pumping loops which permit the use of low-head in-line pumps resulting in reduced horsepower and costs.

Automatic defrost on the air to water heat pumps of the electronic temperature-search type.

Night override provided on the Versatemp circuits so that occupants may run groups of units independently for out-of-hours working.

Figure 9.3 is a diagrammatic schema of the associated air and water circuits.

The building is all-electric in operation with the capability, through the standby generator, of continuing in operation independently of all external sources of supply.

The relevant costs at completion, on a net lettable floor area basis, were:

Installed costs
   All services   £205.5/m²
   Air conditioning £ 99.5/m²
Annual running costs
   All services   £ 7.0/m²
   Air conditioning £ 3.5/m²

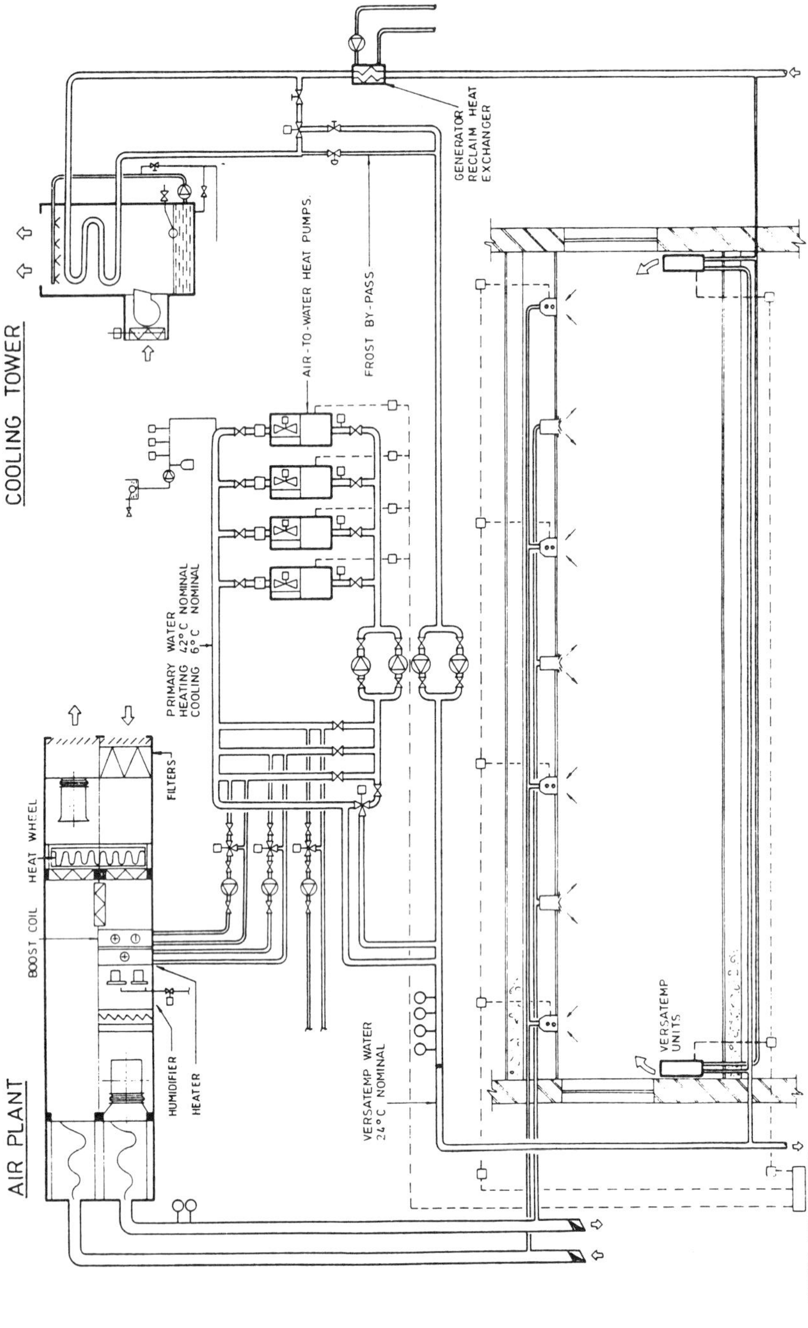

Figure 9.3   *Versatemp system with air to water heat pumps.*

Zoned two-pipe fan coil system with central air to water heat pumps

The terms of reference in this case were for a warm air heating system and, again in 1979, a gas supply was not available and the use of oil was rejected on the grounds of high first costs. Hence, an analysis was made of the economics of using air to water heat pumps; from the point of view that, if viable, the system would provide cooling as a bonus. This proved to be the case and the installation has been running for the past 2½ years in the form shown diagrammatically in Figure 9.4.

This is a simple but effective design which provides heating and cooling simultaneously on a zonal basis. Three 26 kW heat pumps are used, one acting as a top-up machine for either zone when the heating or cooling requirements exceed the capacities of the single machines. Hot or cold water is circulated to ceiling-mounted fan coil units in each zone together with fresh air from a common central air-handling unit. The fresh air is tempered in winter by a speed-controlled thermal wheel recovering heat from the ceiling voids.

The automatic control of the system consists of a changeover thermostat in each zone and individual thermostats operating three-way valves on the fan coil units.

The zone thermostats are set to the nominal design temperatures of the zones above or below which one heat pump will be called in to serve each zone for cooling or heating duty as may be required. Hence, if the thermostats sense continuing deviation from the set point, the third machine will operate as an auxiliary to provide heating or cooling to either of the others, or both.

Individual control to each office is provided by a fan coil thermostat controlling a three-way valve on the water circuit. The operating sense of this control is reversed by the action of the appropriate zone thermostat as it moves from cooling to heating and vice versa.

This is a further example of an all-electric building although, lacking a standby generator, it is not completely self-sufficient as is the case with the first example.

The installed costs at 1983 completion on a net lettable area basis, were:

All services	£194.7/m²
Air conditioning	£ 82.8/m²

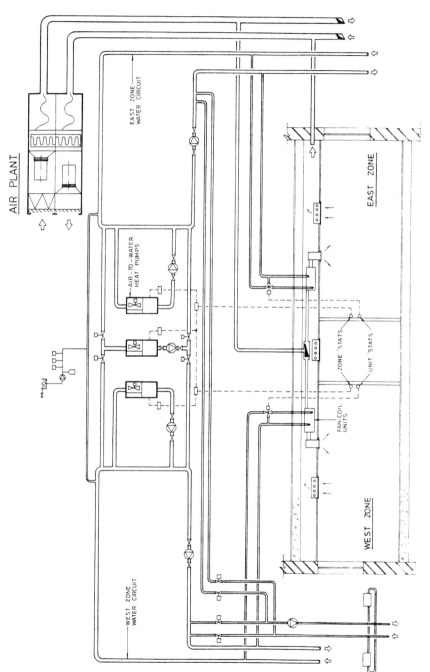

Figure 9.4   *Zoned two-pipe fan coil system with air to water heat pumps.*

Four-pipe fan coil system with air to water heat pumps and auxiliary gas-fired boilers

This is an example of a further form of air conditioning distribution served by packaged air to water heat pumps in place of conventional boiler and refrigeration plant; the system in this case comprising four pipe fan coil units with connecting pipework, ductwork, air diffusers and electrical supplies mounted above the building suspended ceilings.

The luminaires are of ventilated type and these, with the supply air diffusers, provide the suspension for the ceilings in place of the usual system of drop-rods and supporting bars. The ceiling tiles are of lay-in pattern, wholly demountable to give clear access to the units and services above.

The buildings and all services are designed for single or multi-tenancy letting; the latter on a half-floor basis, if required, with provision made for separate metering to each area.

An energy management unit (EMU) allows single tenancies to operate independently of the others in terms of the sequence and duration of use of the lighting, air conditioning, security, etc., relevant to the working times and hours of each occupation.

The EMU also performs the following functions:

- optimizer control of the air conditioning system.
- monitoring of the electrical demand and the shedding of non-essential loads to prevent a preset level of demand being exceeded.
- the switching of the emergency lighting and periodic test by simulated power failure.
- out-of-hours random switching of the lighting for bogus security patrol purposes.
- operation of the car-park lighting through a selenium-cell control.

Figure 9.5 is a diagrammatic schema of the air conditioning system showing the air and water circuits serving the ceiling-mounted fan coil units.

Gas-fired hot water generators are used for heating to the landlord's areas, to two penthouse flats and for hot water service throughout. These generators are also connected to the heating loop of the fan coil units to provide partial standby to the central heat pumps.

Three packaged air to water machines are used to provide heating and cooling to the hot and cold water circuits serving the fan coil units. The machines are programmed to run in any sequence of heating or

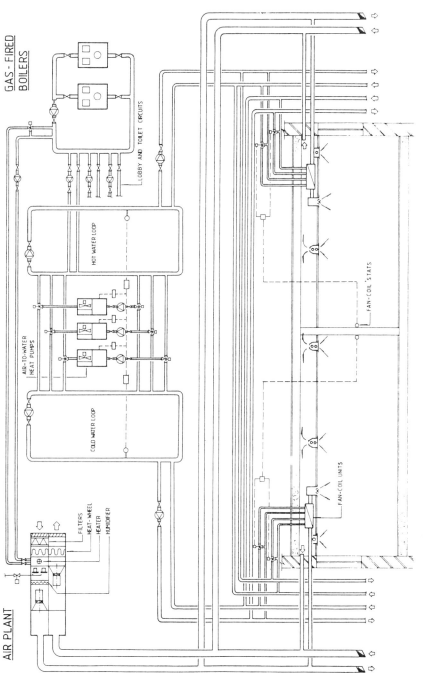

Figure 9.5 *Four-pipe fan coil system with air to water heat pumps and auxiliary gas-fired boilers.*

Figure 9.6   *Typical floor of building with four-pipe fan coil system and air to water heat pumps.*

cooling to meet the prevailing thermal load of the building. Each machine is rated to supply 169 kW of cooling and 136 kW of heating.

Figure 9.6 shows the interior of a typical floor.

Based upon net lettable areas, the relevant costs at November 1983, were:

Installed costs
   All services   £208.7/m²
   Air conditioning   £ 88.8/m²
Running costs
   All services   £ 7.9/m²
   Air conditioning   £ 2.9/m²

Four-pipe fan coil and variable air volume systems served by air to water and water to water heat pumps

This is a recent design for a building at present under construction in the City of London.

Client instructions to the design team were for the building and services to be designed to provide the greatest possible flexibility of use by occupants in a single tenancy, or any multiple of sub-tenancies; to operate with low service charges; and to offer security of tenure in respect of safety, security, operational reliability, etc.

To satisfy these various requirements the building structure has been designed to standards of thermal insulation above those specified by Building Regulations; to be as airtight as possible, and for ease of cleaning and general maintenance.

The services are designed to permit:

- the lighting, power and air conditioning to be supplied, controlled and metered to the building as a whole, or to any combination of sub-tenancy floors.
- to have dual systems of air conditioning for the perimeter and inner zones, served from common central plant.
- to have bivalent energy supplies (gas and electricity) and a standby generator.
- above-ceiling distribution of fixed services (ducts, pipes, cables, etc.) and a raised floor for variable services such as telephones, data systems, task lighting, etc.
- photo-electric control of the installed lighting.

An energy management unit monitors and controls:

- the air conditioning plant.
- general lighting switching, and emergency lighting and test.
- load shedding for maximum demand limitation and for the operation of the standby generator when energized by the EMU on mains failure.
- fire alarm system, bogus security lighting, intruder security, external lighting, metering of tenancy power consumption, etc.

Figure 9.7 depicts diagrammatically the operation of the air-conditioning systems and the heat pumps as controlled by the EMU for maximum running economy.

For as long as conditions will permit at partial load, the thermal demands of the building are met by the air systems and the water to water heat pumps. Thence, as the cooling or heating loads increase, the boilers and the heat pumps are brought into service; the latter as coolers or heaters according to the prevailing load.

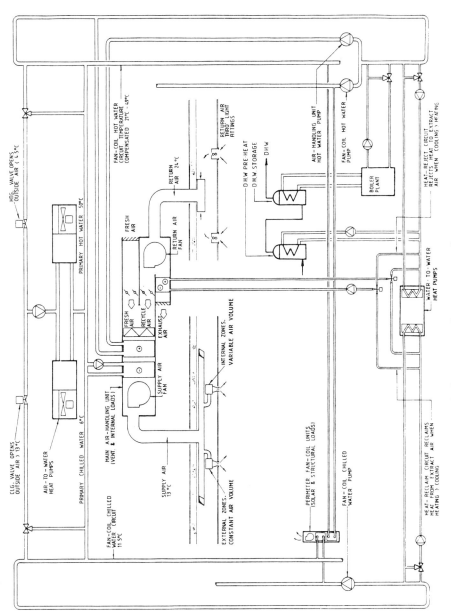

Figure 9.7 Four-pipe fan coil and VAV systems with air to water and water to water heat pumps.

The ratings of the heat pumps are:

air to water machines   3 × 129 kW cooling
                         105 kW heating
water to water machines 2 × 37 kW cooling
                         49 kW heating

In general, the plant operation is divided between three phased conditions of outdoor air temperature, thus:

At and below 4.5°C (40°F)   The supply air plant will run with maximum fresh air to provide free cooling via the main cooling coil. At the same time the run-around water system and the water to water heat pumps assist in the cooling and heating processes, by extracting heat from the building return air and from the fan coil chilled water circuit and transferring it to the fan coil heating circuit.

The free heat so obtained is supplemented by the air to water heat pumps or the gas-fired boilers whichever the EMU selects as being the most efficient in relation to the outdoor temperature.

Between 4.5°C and 13°C (55°F)   Free cooling will continue at the air plant, supplemented as necessary by chilled water through the main cooling coil. The water to water heat pumps and the run-around system continue to operate to provide chilled and heated water to the fan coil cold and hot water circuits.

At 13°C and above   The air plant will operate with minimum fresh air and a greater proportion of return air, the mixture being cooled through the main cooling coil. The water to water heat pumps operate to provide partial cooling of the fan coil cold circuit and the heat required at the fan coil hot water circuit.

The run-around system is used to control the temperature of the hot water circuit by rejecting excess heat to the outdoor air.

The balance of the cooling required in the fan coil cold water circuit is provided by the heat pumps operating as chillers.

Costs at 1983 are as follows:

Installed costs
    All services   £390.5/m²
    Air conditioning   £184.0/m²
Running costs
    All services   £   7.0/m²
    Air conditioning   £   3.4/m²

Conclusion

Over the past 5 years heat pumps in various forms have been used in a great number of air conditioning and warm air heating installations, of which those described are typical examples.

Compared with installations using conventional cooling and heating plant, they have proved to be comparable in first costs and, by calculation and preliminary operating indication, they will produce economies in running costs – particularly when the design of the building and services has been integrated for high energy conservation.

In this respect, two of the installations described above have been examined by computer, in order to compare the annual running costs of the services designed to these standards with those for the same buildings, had earlier, less efficient design procedures been used. The results are given in Table 9.1 on a net floor area basis.

Table 9.1 *Comparison of heat pump designs*

	As designed	*Conventional design*
Building 1		
All services	£7.0/m²	£10.0/m²
Air conditioning	£3.5/m²	£ 5.1/m²
Building 3		
All services	£7.9/m²	£11.2/m²
Air conditioning	£2.9/m²	£ 5.7/m²

Discussion

D. Guy (Denis Guy & Partners)   Where gas is available, has Mr Fowler considered a gas engine heat pump or even private generation of electricity with a gas engine?

K. J. Fowler (Kenneth Fowler & Partners)   These options were not considered and the financial comparison made in the chapter was based on conventional use of gas.

J. Leary (Electricity Council)   Where gas is available heat pumps have a job to be cost-effective for heating only applications.

K. J. Fowler (Kenneth Fowler & Partners)   In our comparisons of gas and oil firing with heat pumps, almost every exercise showed mains gas to be more expensive than heat pumps. However, the building must be designed as a whole with all the services taken into consideration. It is no use comparing the running costs of heat pumps with gas or oil during the winter season only. It must be all-year running, taking into account the cooling and heating plus any other forms of energy conservation being used.

J. Leary (Electricity Council)   My interpretation of the way gas heat pumps are selected for an air conditioning installation is that the compressor is matched with the summer cooling load so that the compressor is equivalent in size to that of a correctly sized heat pump for the same building.

However, the gas turbine drive would supply enough excess heat to heat the building two or three times over in winter time. Therefore, to economically apply gas heat pumps the building should have a heating load to cooling load ratio of about 2:1 which does not apply in building air conditioning.

A gas engine heat pump will normally be cheaper to operate than an electric pump per unit of heat output but there is the problem of over-capacity and the substantially increased capital cost. The capital cost must be an integral part of the considerations. In balancing the cost-effectiveness of energy and capital in air conditioned buildings it can be seen that the building heating energy requirement is far less than the output of a gas heat pump.

B. Dann (South Eastern Gas)   I would endorse what has been said on combined heat and power. The possibility of installing such plant within SEGas buildings has been investigated but as yet a building has not been found that can use the power to heat ratio necessary to make such a unit cost-effective. The main problem is in the heat generated, such that unless a large heat 'sink' is available, as in a swimming pool, then the real cost of the electrical generation rises if the heat produced cannot be effectively utilized.

Editor's note: For additional comments on combined heat and power and heat pumps, see discussion following Chapter 8.

D. Sparks (Staefa Control Systems)   In the air to air four-pipe fan coil system illustrated in Figure 9.5, there is both a chilled water loop and a hot water loop connected to three two-pipe heat pumps. Each heat pump can either heat or cool and is connected to one or other hydraulic circuit through three-way valves, one on the flow to the heat pump and one on the return. Which loop demands the heat pump to run? Is priority given to the chilled water circuit or the hot water circuit? Are ordinary control valves used, for if they are water will be exchanged from the chilled water circuit to the hot water circuit.

In an almost identically designed system, we had problems. During changeover, chilled water flowed into the hot water circuit unless the pumps were stopped. The system was open vented and during changeover, water was pushed up and out of the expansion pipe due to the extra pressure in the hot water circuit.

K. J. Fowler (Kenneth Fowler & Partners)   A sequenced control is used, that is, if the greater demand is for cooling, then some, or all, of the units would operate as chillers; as the demand for heating increases the units would progressively go over to the heat pump cycle.

Being a four-pipe system both hot and cold water are needed for most of the year so both are produced and it is only towards the extremes of the heating or cooling seasons that the heat pumps will go over entirely to cooling or to heating.

R. P. Dearling (Kenneth Fowler & Partners)   The design principle is based on primary and secondary pumping systems, that is: one primary heating pump which circulates water through the heating coils of each fan unit; one primary cooling pump which circulates water through the cooling coils of each fan unit; with each heat pump unit having individual heating/cooling secondary water circulating pumps. The latter are capable of supplying heating or cooling water to the respective primary circuits via the three-port changeover valves. The position of the flow and return entry of the heat pump's secondary circuits are at a point of minimum pressure drop to both primary and secondary systems, ensuring that no water flows through the secondary system during its off cycle. Of the three heat pump units installed, one is designated to supply heated water, one is designated to supply chilled water with the third supplying either cooling or heating water as the seasonal demands vary.

As the heating season ends and the ambient temperature increases, the operation of the fan coil changes from heating to the 'dead zone' mode, which causes the temperature differential of the heating water to decrease, in turn allowing the heat pump unit to shut down in the following sequence.

1   Primary water pump shut down.
2   Heat pump unit shut down, via flow switches and auxiliary pump starter interlock.
3   Power shut down to the three-port changeover valves.

As the demand for cooling increases the temperature differential of the return cooling water increases, thus activating the three-port changeover valve motors which position the valves to the cooling mode. Auxiliary switches fitted to the valve drive motors also position the heat pump unit's reversing valve to its cooling mode and start the secondary water circulating pump serving its heat pump unit. This in turn activates the heat pump unit on cooling mode.

In brief the basic operation is as follows: as heat pump units are satisfied, their respective secondary water circulation pump shuts down, thus ensuring that no water circulates through each unit during the off cycle, and preventing the transfer of either hot or cold water between the primary heating and chilled water circuits.

The installation is now entering its second heating season, and to date no problems of interchange between circuits have been reported.

It has been found, as predicted, that during heating operation one heat pump provides sufficient energy to satisfy the heating demand of the building. This is because maximum energy recovery is incorporated in the design – heat recovery from lighting and people via the thermal wheel installed in the fresh air system; sealed double glazed window units used throughout; and low thermal transmission values of the structural elements. Therefore the number and size of heat pump units were selected to satisfy the cooling demand, which results in an excess heat output capacity, during reverse cycle, to that required by the building.

J. W. Kew (Building Services Research & Information Association)   Was a water to water heat pump considered to transfer heat from one primary circuit to another and further improve the systems energy efficiency of the system?

K. J. Fowler (Kenneth Fowler & Partners)   Although not used in the case just presented in this chapter, a water to water heat pump has been used on a similar subsequent project, and of course water to water heat pumps are used in the fourth case described – it is perhaps a question of evolution. It is as yet too early to give definitive running cost comparisons, but early accounts indicate a worthwhile saving.

G. W. Aylott (Electricity Council)   Mr Fowler has given us a most interesting paper describing a number of different applications of heat pumps. He indicated that in comparison with conventional heating and cooling plants the first costs are very acceptable and the initial experience in the first few years – I think the oldest of the buildings is about 5 years – has been good, but could I ask you what has been your experience (a) with the commissioning of these systems in comparison with conventional systems and (b) in terms of the reliability?

K. J. Fowler (Kenneth Fowler & Partners)   There was some difficulty commissioning the first installation but this was not due so much to the design and equipment used but rather the contractual arrangement. It was a management contract and this gave rise to considerable difficulty with commissioning the job as a whole. Subsequently no more problems have been experienced than with other forms of plant. In fact, fewer problems have occurred because nearly all the equipment was packaged which reduced the requirement for site commissioning and allowed the individual items to be commissioned by the specialist supplying companies.

The control circuitry is perhaps the most difficult to set up accurately; it takes time and requires the attention of expert personnel who work closely with our own control specialists.

It is perhaps a little early to judge the maintenance requirement as the oldest installation has been operating for only 4 years. During that time there has certainly been no trouble with the central heat pumps. Any problems have been with the distribution systems and these would have occurred in any case. Working with packaged, predesigned and tested heat pump equipment, no problems have been experienced to date and none are anticipated.

Dr R. G. H. Watson (Building Research Establishment)   In addition to information on the physical performance of buildings using heat pumps, is there any information about customer reaction to the flexibility of conditions they have created and the views of the people that have used the building?

K. J. Fowler (Kenneth Fowler & Partners)   The answer depends on the type of distribution used with the heat pumps, and also on the office personnel. The first building mentioned in the chapter uses the

'Versatemp' system which is very good but has its limitations, one of which is that the cooling and heating output is provided by a relatively small unit. The staff concerned moved to a new light-weight modern building from one of traditional construction. They were not happy but staff never are when moved from a known location. They were unaccustomed to all the heating and cooling coming from one small unit. The unit controls are such that it is possible to switch from cooling to heating almost immediately and the staff appeared unable to appreciate that the unit control should be left at a set point and the unit allowed to take care of itself.

On arrival in the morning, probably having left the unit control at maximum cooling the previous night, the building appeared cold and the units would be switched right over to heating. An education period is required and the duration depends upon the kind of system being used, particularly when people are moved into a new building. Regarding the other buildings mentioned, I have not heard of any problems but the less obtrusive fan coil system is being used.

10 Heat pump system for the Edmonton Journal building

B. A. Bath

Introduction

The *Edmonton Journal* is a daily newspaper published in the western Canadian city of Edmonton in the oil-rich Province of Alberta. It is distributed to many other towns throughout the northern half of the Province, and with this widespread circulation area, the *Edmonton Journal* has one of the largest daily circulations of all North American newspapers.

In the mid 1970s, as a result of spiralling oil prices, Alberta's economy was booming and with it so was the Province's and hence Edmonton's population which, to some extent, has since proven migratory in nature.

Faced with a daily circulation which, in 1976, already over burdened existing press equipment, and the prospect of circulation doubling by 1990, the *Edmonton Journal* decided to construct a new production facility with increased press capacity and utilizing high-speed, offset process, presses.

Combined with the new high-speed press facility there is a requirement for high-tech support equipment such as laser plate-makers, computerized mailing and sorting equipment, composing and editorial computers, etc.

The newspaper publishing industry is one of the most rapidly changing and modernizing industries with daily deadlines to meet and thousands of newspapers to print everyday, in a relatively short period. To meet these rigid production demands, computerization and utilization of modern communication technologies has become commonplace.

Needless to say, the new equipment would be very heavy in electric energy consumption and very high in heat output, and would require closely controlled environmental conditions. It was a natural progression of the early design concepts to consider methods of capturing or re-using this heat.

As noted, the planned new presses were required to be housed in a closely controlled environment and prescribed conditions by the press manufacturers were 23.8°C ±0.56°C and 50% R.H. ±2% at all times regardless of press operation. These conditions were to be strictly maintained to prevent the build-up of static electricity, which in turn could cause paper breaks, web breaks, equipment failure, etc.

Ancillary areas, such as the paper warehouse were also required to be maintained within pre-set limits of 23.8°C ±1°C and 30±5% R.H. Less stringent, yet still requiring closely controlled environmental systems.

These indoor conditions and precise tolerances, particularly the humidity levels, compared to Edmonton's climate, required the design of sophisticated environmental systems and building construction.

Climate

Edmonton is located at 52 degrees north latitude and 13 degrees west longitude, set on the western fringe of the Canadian Prairies, receiving lots of sunshine and little precipitation.

The Edmonton climate is one of the most severe of major Canadian cities, having a winter design temperature of −38°C (−37°F) with extremes recorded down to −47°C (−53°F). The summer design temperatures are 30°C (86°F) dry bulb and 18.8°C (66°F) wet bulb, with extreme dry bulb temperatures recorded as high as 37.2°C (99°F).

Solar intensity at ground level is high due to very little haze and bright sunshine is frequent. Alberta is known as the 'Sunshine Province' and boasts an average of 2350 hours of bright sunshine yearly.

The heating season is long, totalling 6110°C (11,000°F) degree days calculated on the average daily mean temperature below 18°C (65°F), that is, one day with a mean temperature of −28°C (−20°F) totals 46°C (85°F) degree days. Below freezing temperatures have been recorded in every month of the year.

With these conditions, heating availability is required year round although infrequently utilized during summer months.

The building

The building is a three storey structure enclosing 18,570 m² (200,000 ft²) of gross floor area. Much of the area is a very high, 15 m single storey building with the ground floor comprising 10,820 m² (Figure 10.1).

Figure 10.1   *The* Edmonton Journal *building.*

The building configuration and design emanates from its function as a newspaper production plant and the size and arrangement of space are predicted on the production equipment it houses. The following facilities are included:

1   A basement of 1440 m² housing building mechanical equipment and thermal (water) storage tanks.
2   Ground floor of 10,820 m² housing the reel room (lower press level), paper warehouse, mail room (for newspaper section correlation and bundling), dispatch area and administration offices.
3   A second floor of 3900 m² housing the main press level, locker room, cafeteria and related staff facilities – the press (Figure 10.2) and reel rooms, which are open to each other, for all intents and

Figure 10.2 *One of the four press units being assembled in the 13 m high press room.*

purposes comprise a single space 78 m long, 22 m wide and 15 m high and the entire space has to be maintained within the prescribed temperature and humidity limits without stratification.

4   A third floor of 2410 m² housing mechanical and electrical building equipment and ancillary spaces.

The severe winter climate and rather stringent interior space temperature and humidification requirements dictated construction of a building shell of high thermal quality with particular emphasis on vapour barriers to allow the high humidity levels to be maintained.

Factors influencing heating, ventilating and air conditioning system design

The required tolerance on the building environment led to the design of air conditioning sytems of constant dewpoint control. Rybka, Smith & Ginsler Limited had designed several other projects requiring precise environmental control, including two other newspaper production facilities and one commercial printing plant for the Southam Organization (the owner of the *Edmonton Journal*) and our experience with these projects over the years has illustrated that constant dewpoint control is the most reliable and precise form of air conditioning at a reasonably economical cost. This type of system requires year-round availability of cooling and humidification. Utilizing outdoor air for cooling would have resulted in very high winter humidification costs due to Edmonton's dry climate.

The initially calculated peak electricity demand for the press installation, excluding air conditioning equipment, was 2810 kVA. A summer cooling load of 800 tons was estimated, requiring an additional 800 kVA of electricity at capacity.

Electricity in Alberta is expensive compared with other areas of the country and costs are structured to penalize consumers with short-term peak demands.

It was established in the early stages of design that the electricity cost rate structure, for the proposed *Edmonton Journal* would be $10.00 (approximately £5.00) per kilowatt monthly demand charge with the first 500 kWh per kilowatt of demand included in this charge. The estimated demand of 2810 kVA, therefore, would allow 2810 × 500 = 1,405,000 kWh of electricity consumption at no additional charge to the demand. The estimated monthly consumption was 950,000 kWh and it was obvious that electric energy should be utilized at least to the extent of 1,405,000 kWh per month before any other fuel sources were considered, even though natural gas was very inexpensive ($1.00 per MCF/5p. per therm) at that time.

From this it was obvious that anything that could be done to reduce peak electrical demands would yield substantial savings.

An area where substantial reduction to peak electricity demands could be achieved was the air conditioning system. The estimated load of 2800 kW would cost $8000.00 (£4000.00) per month if refrigeration equipment operated coincident with the newspaper presses.

In order to avoid coincident electrical demands for chillers and presses, storage of cooling energy was required. Water was chosen as the storage medium as this could be utilized in conjunction with heat recovery chillers (heat pumps) to capture excess press heat while reducing peak electricity demand, and a system was designed to allow cooling to be stored at any time that the presses were not operating or that building electricity demand was less than the previous peak of that month.

Building load profiles (electricity)

As noted above, the initially estimated peak electricity demand, excluding air conditioning, was 2810 kW, which is illustrated in Figures 10.3 and 10.4, and comprises the following:

* lighting   – 330 kW
* presses   – 1700 kW
* heating and ventilation   – 480 kW
* production equipment (mail room, etc.) – 300 kW

The estimated peak cooling load for the building was 2800 kW, imposing an additional 800 kW of electrical demand which, without thermal storage, would be added to the building loads as illustrated in Figure 10.5.

With thermal storage provided, the peak demand is limited to the requirements predicated by the production of newspapers, as illustrated in Figure 10.4.

These figures were used in the early systems analyses; however, later during the finalization of system designs, other similar newspaper

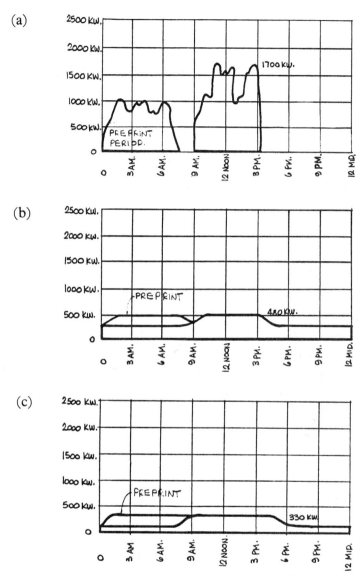

Figure 10.3   *Profiles of electrical demand. (a) Presses. (b) Ventilation systems.*
(c) Lighting.

production plants were metered (electrically) and it was found that
more diversity than originally anticipated could be applied to the
presses and the building peak demand estimate was reduced from
2810 kW to 2340 kW.

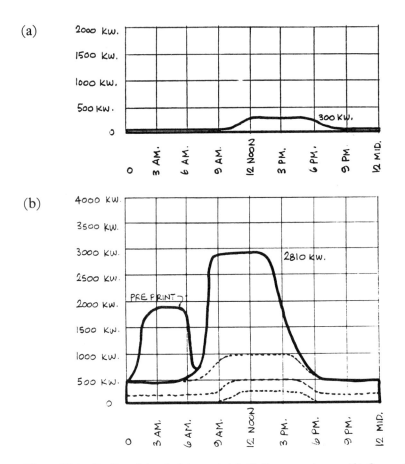

Figure 10.4   *Profiles of electrical demand. (a) Mail room equipment. (b) Composite.*

The system concept

A building environmental system was designed utilizing heat pumps
with thermal storage, with particular emphasis given to energy conservation and operating efficiency. The system is illustrated schematically
on the heating–cooling system flow diagram (Figure 10.6).

The primary component of the design is the central heat recovery
refrigeration plant consisting of one 'standard' and two 'heat pump'
refrigeration machines.

The refrigeration plant operates in unison with a large (1900 m³)
water storage tank. The tank consists of four separate sections each

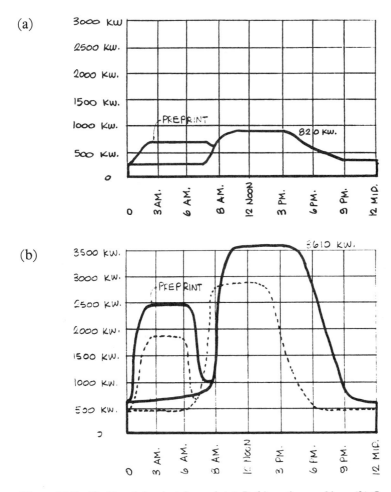

Figure 10.5 *Profiles of electrical demand. (a) Refrigeration machines. (b) Composite of building loads plus refrigeration.*

fully compartmentalized and fitted with flow diverters to achieve the highest possible temperature stratification for optimum performance.

The dynamic characteristics of the energy reservoir (tank) were computed with a building simulation computer program which indicated a 77% thermal efficiency of the tank although 65% efficiency was assumed in the system design.

The tank was designed to store chilled water with the primary operation of the system being the daily storage of chilled water for use during press run time to enable the press heat to be removed from the

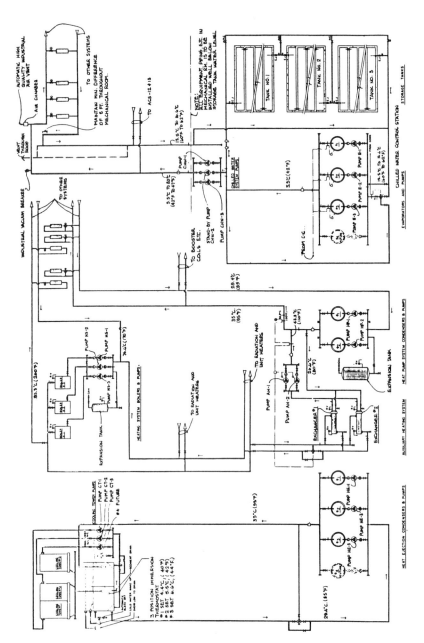

Figure 10.6 Heating–cooling system flow diagram.

press room without simultaneous operation of refrigeration machines and presses.

The hydraulic design of the system allows simultaneous use and storage of chilled water through use of variable speed pumps on the secondary chilled water loop. These pumps provide a varying chilled water flow to satisfy varying building cooling loads while the constant speed primary chilled water pumps, coupled to the chillers, attempt to satisfy the storage requirements under control of the appropriate computerized program.

The heat that is extracted from the press room and other building areas is recovered and returned to the storage tanks as warmed chilled water and is utilized for building heating through the heat pump refrigeration plant, during off-peak times, for building heat, fresh air preheating, and humidification.

In essence the heat produced by the presses and other equipment, for which the owners have already paid, is captured, stored and re-used.

A gas-fired boiler plant is provided to augment the heat pump system during extremely cold weather. The same unit heats water and provides heating to areas requiring a higher than normal temperature heating medium.

The overall system is fully automated through a computerized control console. A separate, highly sophisticated program is used to measure cooling and heating potential within the storage tank. Refrigeration machines are activated automatically to meet daily heating and cooling demands related to outside temperatures and press run characteristics.

The installation has great flexibility. Space is provided for additional facilities to serve expanded press equipment, as needed, and the potential for receiving and processing other heat sources available 'free' or at nominal cost is virtually unlimited. Some examples are given below:

- solar energy
- heat from lights and people (included in the implemented design)
- heat recovery coils in all major exhaust outlets
- heat available in boiler room and electrical room
- heat recovered from boiler chimneys

The important aspects of the system are the cost savings realized in the heating season through re-using the presses and machine-generated heat, and during the cooling season, by the substantially reduced electrical peak demand charges. It was estimated that the additional costs entailed in installing the *Edmonton Journal*'s 'heat pump' system instead of a conventional 'free cooling' design would be recovered in 2–3 years as indicated in the following cost comparison:

1   Cost of additional equipment for a 'conventional' heating, ventilation and air conditioning system utilizing outside air for 'free' cooling when outside temperatures permit:
    a   additional refrigeration plant – $70,000.00;
    b   additional steam boiler plant for humidification – $90,000.00;
    c   additional electrical transformers, switchgear and distribution – $50,000.00;
    representing a total additional cost of $210,000.00 over the designed system.

2   Costs of additional equipment for the designed system utilizing thermal storage and heat pumps:
    a   water storage tanks – $250,000.00;
    b   additional controls including central computer – $100,000.00;
    c   double bundle condensers (heat pumps) – $30,000.00;
    representing a total additional cost of $380,000.00 or $170,000.00 net extra over the conventional system.

The calculated annual energy saving for the heat pump system was $64,000.00, comprising $31,000.00 electricity costs, $13,000.00 humidification costs, and $20,000.00 heating costs.

The feasibility of a heat pump system lies in the availability of an adequate heat source and having a use for recovered heat; the *Edmonton Journal* has both.

System operation

A central control and monitoring system was provided to ensure that system operation was as designed. Several operational computer programs were developed to control the heat pump/thermal storage system to ensure that electric energy demands were kept to the absolute minimum and that sufficient cooling energy was stored in the tanks to carry the building through a scheduled press run with minimum refrigeration plant operation.

These programs monitor at all times the amount of cooling energy stored within the tank, at either the summer or winter operating temperture, and calculate the storage rate required to meet the next scheduled press run and compares the actual storage flow rate with the calculated required flow to satisfy the above, and makes the information available to a 'charging' programme.

The programme facilitates various modes of operation including charging mode, extra press run mode, using mode (summer operation), using mode (winter operation) and reverting to 'storing' mode (summer and winter).

With reference to the system schematics, the following more detailed operational description can be followed.

Storing mode (Figures 10.7 and 10.8)

In the storing mode, the central control and monitoring system program determines the energy required to re-charge the chilled water storage tanks by a pre-set time, the time available, and then calculates the rate of storing required to meet the deadline.

The program continuously compares the flow rate required to recharge the tanks with actual flow rate, and if the required flow exceeds the actual flow by more than 28.4 litre/s, the next chiller is started. If the difference between the required flow rate and the actual flow rate exceeds 56.8 litre/s, another chiller is started. If the required flow is less than the actual flow with one or more machines operating, the entering chilled water setpoint is increased to reduce the bypass around the chillers.

If actual flow exceeds the required flow, the program increases the amount of flow bypassed by reducing the chiller entering water setpoint, thus unloading the chillers. If the bypass flow exceeds 28.4 litre/s, one chiller will stop.

If the chiller entering water temperature exceeds the chiller design conditions of 15.6°C in summer and 18.4°C in winter, an additional machine starts to allow for increased bypass. If the actual flow is too low, and the setpoint has been gradually stepped up to 95% of the operating chillers without obtaining the required charging flow rate, then an additional chiller starts.

The heat recovery machines are allowed to load to at least that as determined by the heat recovery heating loop demand; however, chiller start-up and shut-down are controlled by the charging flow requirement.

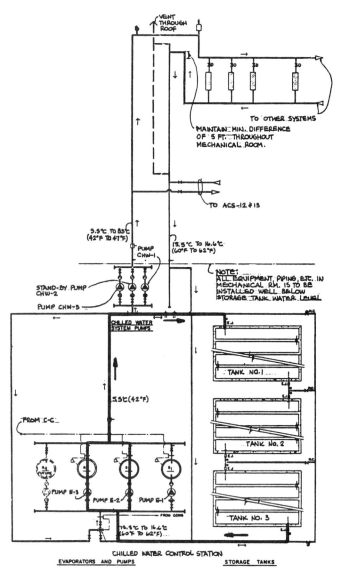

Figure 10.7 *Storing mode with no building demand.*

For example, start-up could be required because:

1 actual flow is too low and the central control and monitoring system is asking for a 95% setpoint on chillers operating, or
2 actual flow is more than 28.4 litre/s below the required flow rate, or
3 chiller entering chilled water temperature is above design temperature.

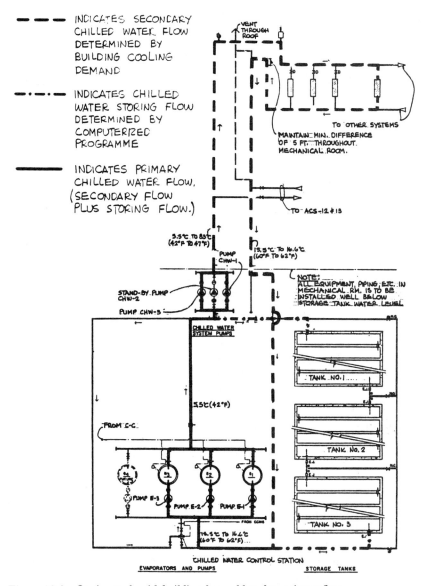

Figure 10.8 *Storing mode with building demand less than primary flow.*

Shut-down could be required because the bypass flow exceeds 28.4 litre/s.

Automatic chiller start-up will always be delayed 1 minute, and chiller shut-down will always be delayed 5 minutes to verify reading stability (period will be adjustable).

Extra press run mode

When the 'extra press run' mode is manually selected, the chillers are loaded to supply a charging flow rate which is greater than the calculated required flow rate by as much as twice the required load. The amount of extra chilled water generated is adjustable. The 'extra press run mode' is automatically shut-off at a preset time.

Using mode – summer and winter (Figures 10.9 and 10.10)

When in the 'storing' mode the system attempts to fully charge the storage tanks by a preset time; however, should the tanks not be fully charged by this time, or due to building load the chillers are still operating at this time, then no changes will be made and the programme will remain in the storing operation. As soon as one of the presses is started, the system changes to the using mode and operates as outlined below.

Summer operation

All chillers are stopped. The lead heat recovery chiller and the summer chiller will be allowed on-line if required. The lag heat recovery chiller is off-line.

If all machines are off and the temperature in the last tank cell or any other selected cell exceeds 42°F, the lead heat recovery chiller will start and the heat recovery loop temperature controller will load the machine as required to satisfy the heating demand.

If the lead heat recovery chiller is on and the energy stored in the tanks is reduced to 20% of storage capacity, and the summer chiller is off, the summer chiller starts after a preset time delay. The summer chiller shuts down when the tank charge returns to 95%.

The lead heat recovery chiller is shut-down by its own load control when its load is reduced to the point where the chiller operates on hot gas bypass.

If building electrical energy demand increases such that the previous peak will be exceeded, then the chillers are unloaded to 40% capacity or turned off by the program as required. Chillers are allowed to re-start only after a preset period.

Chilled water produced in excess of instantaneous load requirements

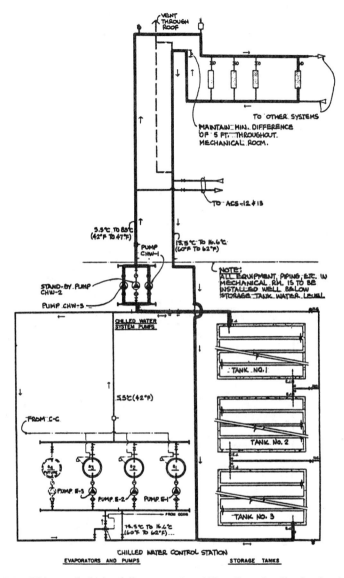

Figure 10.9   *Using mode during full press run (no chiller allowed on line by electrical demand limiter).*

is stored in the tanks; however, if the building load exceeds the chilled water produced by the chillers allowed on-line, then the balance of water is automatically drawn from the tanks. No hot gas bypass operation of the chillers is allowed while operating in this mode.

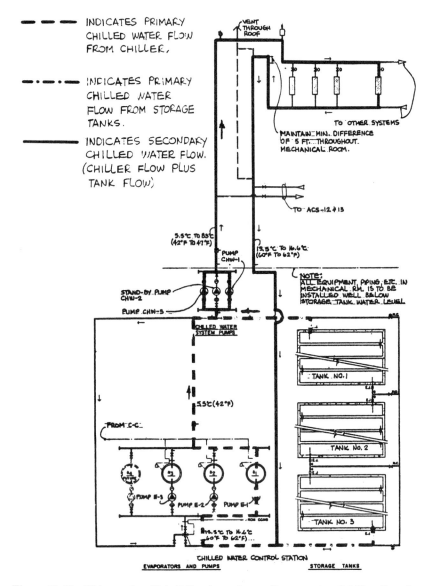

Figure 10.10 *Using mode with building demand exceeding capacity of chiller allowed on by electrical demand limit.*

Winter operation

The lag heat recovery chiller and the summer chiller are shut-off. The lead heat recovery chiller only is allowed on.

If all machines are off and the temperature in the last tank cell or any other selected cell exceeds 14°C, the lead heat recovery chiller starts after a preset time delay. Chiller loading is done from the heat recovery heating loop to satisfy heating demand.

If this chiller is running, and the temperature in the last tank cell or other selected cells falls below 10°C, the machine is shut-down.

If the building electrical energy demand approaches the previously set peak, the chiller unloads or shuts-down as required. The chiller is allowed to re-start only after a preset period.

No hot gas bypass chiller operation is allowed, hence the lead heat recovery chiller may be shut-down by its own controls when it reaches the hot gas bypass position.

Reverting to 'storing' mode (summer and winter)

Thirty minutes after all presses have been shut-down, the system automatically reverts to the storing mode and immediately calculates the flow required to re-charge the tanks by the preset time and attempts to obtain the required flow rate.

In the winter, the flow rate will be selected to prolong the storing hours, but in the summer the storing rate will be at full capacity during the day and at normal calculated capacity during the night.

Demand limits

At any time, summer or winter, regardless of mode, the chillers and other listed electrical equipment are unloaded and/or turned off whenever the building electrical demand approaches the previously set demand limit. This programme overrides *all other* programmes.

Operational experience and actual energy consumption

The system was commissioned in 1980 with little difficulty.

Turbulent flow was experienced in the return piping when primary chilled water pumps were operating, which caused malfunctions. The cause was found to be air being drawn in through the riser vent. The problem was corrected with minor piping revisions.

Another minor difficulty was experienced from return chilled water being drawn through off-line refrigeration machines by the secondary chilled water pumps at certain flow conditions, causing higher than design water temperatures. This was overcome by automating a shut-off valve on each chiller and control revisions to close the valve when its refrigeration machine was off.

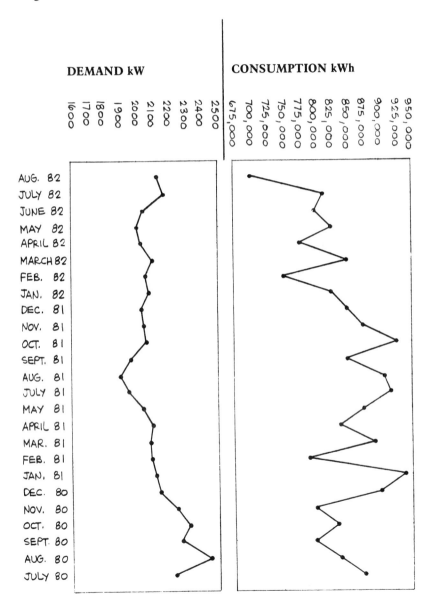

Figure 10.11   *Energy consumption graph.*

Initial energy costs were high due to press commissioning and mal-functions in the central control and monitoring system; however, after a 2 or 3 month debugging period the energy consumption was reduced and actual consumption is slightly less than design forecasts. The actual energy consumption from July 1980 to August 1982 is indicated on the energy consumption graph (Figure 10.11).

Discussion

R. B. Watson (Davis Belfield & Everest)   Mr Bath mentioned a 3-year pay-back period in relation to the *Edmonton Journal* building. He also said that Edmonton is a Canadian city with one of the most severe climatic conditions. To what extent is the 3-year period dependent on the energy requirements arising from these severe climatic conditions?

B. A. Bath (Rybka, Smith & Ginsler, Ontario)   For this particular system, the *Edmonton Journal*, the climate did not play a big role in the cost analysis. The major contributive factor was the charges levied by the power companies, and the design was implemented more for energy cost savings than purely energy savings. It is simply that the electricity rate structure in Edmonton permitted us to save the client an enormous amount of money by peak load shedding and thermal storage.

G. Braham (Electricity Council)   Thermal wheels and heat exchangers are common features in all the heating/cooling systems described in Chapter 9. What are the advantages of thermal wheels or other forms of heat exchanger in building ventilation systems?

B. A. Bath (Rybka, Smith & Ginsler, Ontario)   In the *Edmonton Journal* building it was not necessary to recover any more heat than that recovered from the press equipment. In other systems designed for buildings without such tremendous internal heat gains, heat recovery devices have been used. Our preference is to stay away from heat wheels; we have had significant problems with them over the years. They have been around since the fifties and are well proven, but still we have found that building operators play with speed controls, filters do not get changed regularly and the wheels get plugged up, etc., which affects their performance. Of course, they are fairly efficient devices when properly maintained. In more recent years, static plate-type heat exchangers have been preferred. These are to all intents and purposes maintenance free, apart from systems with defrost cycles incorporated on them. So our preference has been for static plate exchangers where we needed to regain more heat for conventional design, sacrificing a few percentage points of efficiency but gaining in maintenance-free operation.

J. Leary (Electricity Council)   In simple changeover systems with ventilation duct work distributing fresh air, the only way of extracting heat from the exhaust is by using a recuperator or a thermal wheel and it is worth doing. In a more complex system distributing heating and cooling throughout the building, I much prefer to recover exhaust heat by the refrigeration machine and recycle it throughout the whole system rather than offer it solely to the incoming fresh air. The reason for this is so that the free cooling effect of the fresh air can be utilized in the system as much as possible. Even with the simple system, if heat is recovered and used to warm the incoming air on a changeover system with a simple fresh air duct, the free cooling effect needed to make the system work is lost. With a changeover system on heating cycle the fresh air is relied upon to provide the cooling and therefore that air should not go above 12°C. It is likely to be only in the depth of winter that the heat wheel or recuperator will be used, the rest of the time it will be bypassed.

E. J. Bennett (Metal Box plc)   At Metal Box, with an energy expenditure of some £15 million per year, we are always looking for means of recovering energy and heat. Thermal wheels have been used, not directly for space ventilation heat recovery but for heat recovery from printing and lacquering ovens. This might be said to be a poor example of the use of a thermal wheel bearing in mind the solvents and solids given off in these exhaust systems. However, we have found that the thermal wheel in one of our factories within 3 months becomes heavily coated with lacquer, but nevertheless still performs thermally in the region of 70–75% heat recovery compared to 85% for a clean wheel; there is very little fall off. The other advantage of the thermal wheel is that it picks up the latent heat in the exhaust air.

The matrix is changed yearly being replaced by a clean one. The dirty stainless wire matrix is then at our convenience ultrasonically cleaned. This procedure represents a saving of approximately £8,000 per year in energy. The heat wheels are in use continuously 24 hours a day, 365 days a year. Plate-type recuperators have been used at various other locations, both at low and high temperatures, and the return on these is about 65%. The limitations on both these types of unit, particularly if they are retrofit, is the location of supply and exhaust ducts.

Run-around coils have a lot of advantages in terms of heat recovery, particularly if retrofit, because exhaust ducts and inlet ducts can easily be coupled together via piping and a return of about 50% is typical.

J. Leary (Electricity Council)   If heat recovery wheels have a good load factor, pay-back is fairly quick. In many office buildings with ventilation, pay-back is rather extended and filters are needed to stop the wheel becoming clogged. With low load factor and the problem of clogging up, is the expense justified?

D. Sparks (Staefa Control Systems)   Heat recovery wheels can be operated very efficiently when linked with a control system. Alternatively, on general HVAC systems, monitoring the air quality in the space or extract air could be used to control and therefore reduce fresh air input which would reduce energy loss through the exhaust system. Control is based on monitoring contaminants such as CO_2. Reduction in fresh air levels could in some cases be reduced to zero. Such a system would require simple modulation of the fresh air and extract air dampers.

B. A. Bath (Rybka, Smith & Ginsler, Ontario)   I would caution on the reduction of fresh air as a means of saving energy; this has been done in North America over the past 15 years and many buildings now are paying the price. There is a syndrome known today as the 'sick building syndrome' where fresh air quantities of compartmentalized systems were designed to preset *minima* around 4.8 litre/s per person, and many buildings have been operating for a number of years with this limited amount of fresh air. What is being found now is that chemical emissions, such as formaldehyde from office furniture, and other contaminants are building up in the building, creating a 'sick building'. North American buildings tend to be very well sealed in terms of building skin; some buildings in the UK are less well sealed, and vapour barriers are not as important as in the colder regions of North America, so I would say that, if anyone proposes reducing fresh air as a means of energy conservation, it should be looked at extremely carefully. Millions of dollars are being spent on some very large buildings in Canada to retrofit systems to provide more fresh air, and some of these buildings are only 8–10 years old. With conventional systems, where outside air is used for free cooling, the thought is that when on a free cooling mode the building is being purged of the contaminants, but when systems are designed with minimum fresh air, where only 10–15% can ever be put in, there is no purge and the contaminant levels increase.

D. Sparks (Staefa Control Systems)   Fresh air content control by space contaminant sensing is an up and coming idea currently being tested. A number of different sensors are available such as a CO_2 detector, and detectors that generally sense combustible gases, the type of contaminants given off by plastics and other products. This would help overcome the problems previously mentioned in North America.

The use of contamination sensing, and modulating fresh air dampers (in other words fresh air content), was described recently on British Television.

B. A. Bath (Rybka, Smith & Ginsler, Ontario)   Obviously, if all the contaminant levels in the building can be monitored, by all means reduce the fresh air; but there may be a lot of contaminants from synthetic materials and fibres, etc. about which we are not aware. We know there is a problem in some North American buildings, probably more in North America than here because buildings are maintained under positive pressure for the most part, so we do not have infiltration, just exfiltration. Experience with buildings here is that they are less tight and usually under negative pressure.

K. J. Fowler (Kenneth Fowler & Partners)   If a heat recovery device is not being used a substantial form of recovery from lights and people, etc. is being missed and recovery is quite easy using a thermal wheel.

J. Leary (Electricity Council)   The minimum fresh air means to me sufficient but no more than is sufficient and recognition of what is sufficient fresh air is important. I see the minimum fresh air as that specified in the CIBSE guide for the particular application. I certainly would not advocate going below that level under any circumstances, certainly not to save energy. The question 'Why should the fresh air volume be increased above the guide figures?' is an important one.

B. A. Bath (Rybka, Smith & Ginsler, Ontario)   The levels which are recommended in the guides, CIBSE or ASHRAE, are based on odour control and respiratory requirements in the building, to keep it fresh and to prevent people from falling asleep in the middle of the afternoon from lack of fresh air. These guidelines were set many years ago, before synthetic materials were so widely used in office furniture, etc., and the word of caution is not that minimum fresh air is not sufficient to support life but that it may not be sufficient to purge the building of unknown contaminants.

D. Sparks (Staefa Control Systems) It is very difficult to control the fresh air content with variable volume systems because of the requirement to set the fresh air damper positions to the maximum room air volume condition. When the load reduces, and the variable volume system turns down to minimum volume, the fresh air intake damper is still in the same position as that required for maximum volume and far too much fresh air is introduced into the building. There are methods of controlling the fresh air requirement other than using contamination sensors. The use of a fresh air velocity sensor to measure the fresh air intake volume and override the damper position is an alternative.

B. A. Bath (Rybka, Smith & Ginsler, Ontario) The way to overcome that is to put in a small fresh air fan and control the fresh air independently of the mixed air damper.

J. Leary (Electricity Council) It is important not to over-reach the fresh air provision. The more air put in in winter, the lower the humidity, which is a hazard in its own right and may be considered a greater discomfort than contaminants, apart from the problem of static electricity. If a VAV system turns down at minimum cooling load the system will still be fully occupied.

R. Godoy (ECD Partnership) We are monitoring the project mentioned by Mr Sparks. In that experiment the amount of fresh air provided per person is kept constant and at the value given by the local regulations or recommendations in, say, the CIBSE guide. The total fresh air volume is varied with the number of occupants in the building, so that the amount of fresh air is reduced when fewer people are in the building. The variable ventilation system is controlled by a CO_2 detector in the exhaust air ductwork, and may result in significant energy savings in buildings.

D. Terry (North Thames Gas) Heat pipes can operate at very close temperature differential between the supply in the exhaust. They would appear potentially to provide a more efficient form of heat transfer than thermal wheels, etc. Has anyone experience of using these devices?

B. A. Bath (Rybka, Smith & Ginsler, Ontario)   We have used heat pipes to a limited degree. In endorsing static plate exchangers there was no intended implication that it is the only method of heat recovery we would use.

There are many applications where the run-around cycle has advantages, it certainly has gained favour in retrofit and also in some new installations.

The heat pipe has a value for specific applications. They are fairly costly compared with run-around coils and depending on size can be more costly than an equivalent static plate heat exchanger. Cost therefore becomes a factor as well as energy efficiency.

D. Murphy (Kodak Ltd)   Five or six years ago ideas of using ice banks in commercial buildings to reduce air conditioning plant costs did not get very far on cost grounds. The use of storage in Mr Bath's scheme was therefore of special interest. With a greater push on off-peak electricity this could be an appropriate time to develop applications of thermal storage.

J. Leary (Electricity Council)   When using a heat pump, the installed cooling capacity is all used for heating in the winter with an additional requirement for supplementary heating so there is no particular pressure to reduce it. There is no need therefore to store cooling in such a system which represents 99% of installations and where the heat source is external to the building. If there is a great deal of internal heat to recover storage may well be a valid concept. In deep plan buildings I have investigated over the years, there has never been sufficient internal heat to recover to store to warrant the expense. The heat is nearly all used at the time it is recovered. In a building which does generate a lot of heat, perhaps a large computer installation, there may be no way to use stored heat and a heat sink would be needed. This balance is always a problem. The overall cost of storing cooling energy in ice banks including the tank and the controls is expensive whereas cooling capacity is not that expensive. I have always found that the cost of storing energy is greater than the cost of the equivalent capacity. Cooling needs a very high storage capacity for a very short period of time. In the U K it is very difficult economically to justify storage of either heat or cooling and this is particularly so with heat pump systems where the cooling capacity can be changed over to heating.

B. A. Bath (Rybka, Smith & Ginsler, Ontario)   In *The Journal* design, although we store chilled water, we in fact are storing heat while using this chilled water. We then use the warm chilled water to load the heat pump at night and get usable condenser heat out of the chillers.

J. W. Kew (Building Services Research & Information Association)   Mr Bath indicated he was under the impression that the main use of heat pumps in the UK was for space heating only and earlier chapters might confirm that impression.

German experiences with heating only heat pumps were described in Chapter 7 by Dr Paul. It appears that this sector of the German market has now collapsed. At the first Heat Pumps for Buildings Conference* 18 months ago many of us were feeling that we should be following the successful German use of heating only heat pumps. Thankfully we did not follow them on this occasion.

1   Sharma and Dann (Chapter 5) described a heating only gas engine driven heat pump, a £400 saving with additonal capital costs of about £25,000.
2   Chapter 4, by Gregory indicated the difficulties found in matching the heating output of heating only heat pumps to the load.
3   Mr Watson (Chapter 6) told us that there was only limited information on running costs and capital costs of heat pumps.

All this led to a fairly depressing viewpoint that the heat pump was dying and was perhaps dead according to Dr Paul (Discussion, Chapter 7).

Chapters 8–10 have demonstrated the living side of heat pumps, the heating and cooling aspect is a very important one and I would rate it the best application along with the use of heat pumps for dehumidification applications in swimming pools; both very good proven applications.

Third in the hierarchy after dehumidification and air conditioning I would put heat recovery applications. There are demonstration schemes under way at present which will show very interesting results over the next few years for the viability of various heat recovery

* Sherratt, A. F. C. (1984) (Ed.) *Heat Pumps for Buildings*, Hutchinson, London.

applications, particularly hot air exhaust, for example in kitchens and laundries. It is felt that there are problems with grease in kitchen and lint in laundry exhausts. The information coming through suggests we should not be quite so pessimistic about these problems and there is time yet for improvements and developments to be made in these demonstration projects.

Mr Cross of British Telecom asked the question (Chapter 4, Discussion) 'should I use a heat pump?' He was given a reply with a qualified yes, if he was considering a heating and cooling application and we stressed in Chapter 4 that there should be a real need for cooling. If you have that need, heating and cooling from the same piece of equipment, a heat pump is an attractive proposition in capital and running-cost terms. In the earlier chapters all the other applications of heat pumps seemed to be dismissed far too readily.

To summarize (there are many different applications of heat pumps) my recent work at BSRIA identified 15 broad groupings for building services applications alone; ranging from very good to very poor.*

Chapters 8, 9 and 10 provide an excellent discussion of the best applications of heat pumps and provide a positive pointer to the future.

K. J. Fowler (Kenneth Fowler & Partners)   A cooling application is not essential. The second job described in Chapter 9 was a warm-air heating system for which a heat pump system was used because it was cheaper to install than oil and possibly also gas and is cheaper to run. The cooling is a bonus. The annual running costs will be higher in practice than for a heating only system because of the use for cooling purposes.

* Kew, J. W. (1984) *Heat Pumps for Building Services*. BSRIA Project Report, July 1984.

11 Heat storage: realities and possibilities

T. R. Buick, R. R. Cohen, P. W. O'Callaghan and S. D. Probert

Energy and economics

Some 30% of the United Kingdom's consumption of primary energy is used for space heating of domestic, commercial, and industrial buildings. Many opportunities exist for improvements in the efficiency of the means by which this heat is supplied and released. One of the major contributions to reducing this country's consumption of primary energy can be achieved by the use of heat pumps to 'generate' the heat required for space heating. So far, heat pumps have not gained general acceptance as alternatives to conventional fossil-fuel fired boilers because of the higher associated capital and maintenance costs, questionable reliabilities, long pay-back periods, noise generation and poor performances during cold weather.

The heat pump is unique in that it is able to up-grade low temperature energy, which would otherwise be wasted, to a higher and more useful temperature. Many possible heat pump sources and sinks are localized, and subject to cyclic variations in thermodynamic availability, which reduce their usefulness to a simple heat pump. A thermal energy store may be used to control the temporal fluctuations of such sources and sinks, enabling the maximum possible seasonal COP to be achieved, associated with a reduction in the rated capacity of the heat pump.

A further benefit accrues if the high-grade energy tariff exhibits a cyclic cost structure which is out of phase with the heat demand, for example, as occurs with 'off-peak' electricity. Thermal energy storage (TES) can enable the heat pump to make use of the reduced tariff, resulting in significant reductions in the running costs. A thermal energy store could also be used to reduce peak demand, and therefore might offer a similar benefit for consumers who pay a high peak-demand tariff. Where neither of these options exists, the inclusion of storage must be justified on grounds of improved efficiency, and/or reduction of capacity.

Thermal energy storage

Heat can be stored by raising a material's temperature (sensible heat storage), changing its phase (latent heat storage), or separating some of its chemical constituents (thermochemical storage).

Sensible heat storage

Solids

Common examples are night storage heaters using off-peak electricity, pebble beds in solar systems, and ceramic regenerators for industrial waste-heat recovery. The heat transfer fluid is often air, so large temperature swings are possible, and indeed important, to compensate for the low specific heat capacity of solids. The fabric of buildings can also be considered in this category.

Liquids

Generally water or a water/anti-freeze mixture is favoured for temperatures up to 100°C, and heat transfer oils for higher temperatures. Water with its ubiquitous availability, benign nature, and high specific heat capacity is the predominant material for heat storage. Its containment can be sub-surface, for example aquifers or caverns, or above ground, where it can be further classified according to storage volume. Tanks for storing water can be constructed of plastic, metal or concrete and are typically termed small, medium and large for volumes up to 10 m³, 10–100 m³, and 100–100,000 m³ respectively. Larger stores exist in the form of reservoirs, either natural or man-made. A special example of this kind of store is the solar pond, with its relatively efficient means of regeneration.

Gases

This category includes steam accumulators, and advanced compressed-air storage systems.

Latent heat storage

Solid–liquid

This is the most convenient change of phase to exploit. Volume altera-
tions during the change of state are relatively small (up to about 10%),
and latent heats are high. For applications around 0°C, the ice–water
transition is expedient. For higher temperatures, up to 100°C, organic
materials such as paraffin waxes or inorganic materials such as salt
hydrates are available. The organics are in general more reliable, but
have lower densities and therefore lower volumetric heat capacities than
the inorganics. Salt hydrates often require nucleating agents to reduce
super-cooling, and, if they are incongruent, some means of preventing
segregation. The latent heats and melting points of some typical materials are shown in Figure 11.1.

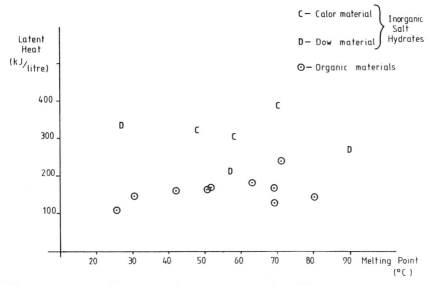

Figure 11.1   *Latent heats and melting points of some typical TES materials.*

Solid–solid

Certain materials, for example memory metals such as nitinol, change
their state when heated through a certain temperature, without changing phase. However, their latent heats are generally small, and the
metals themselves are highly expensive, so that any operational advantages of a single-phase system are usually heavily outweighed.

Liquid–vapour

These systems are dominated by the water to water vapour transition, either in conventional steam to condensate systems, or in association with hygroscopic materials such as silica gel. The latent heat is very high, but storage of the vapour phase can be expensive, or bulky. With hygroscopic materials, the vapour is usually released to the surrounding air. Recovery of the vapour from the environment may cause too much dehumidification, for example in ventilation air.

Thermochemical storage

Solid–vapour

These systems comprise a solid absorber and a working fluid. Probably the most advanced system is being developed in Sweden, and is known as 'TEPIDUS'. It uses sodium sulphide and water. Both the latent heat of the reaction and the latent heat of vaporization of water are effectively stored: the latter is ideal for long-term shorage, that is, weeks or months, as it does not need to be maintained by insulation.

Liquid–vapour

These systems work on the same principles as the solid–vapour ones. The most commonly researched system uses sulphuric acid and water. It is a common feature of thermochemical storage that it involves at least one highly reactive substance which, almost by definition, is toxic, flammable, or corrosive. There is therefore doubt about the acceptability of such systems for the domestic environment.

Heat transfer and heat storage

Effective heat transfer to a heat store is necessary for it to be a worthwhile component of a system. In special cases, for example for many water stores, the storage material is also the heat transfer fluid. In such cases, the thermal performance of the system is influenced by the degree of stratification or mixing that occurs in the store. However, the effect is highly application-dependent. Heat stores containing solid storage materials form another category which has inherently simple heat transfer mechanisms. The heat transfer fluid, often air or water, flows through the store in direct contact with the storage material. Although simple, the quality of heat transfer is often low and depends upon the flow rate, conductivity of the storage material, etc., and for gas-based systems, the pressure drop can be high. One exception to the above is storage in the ground. In this case a heat exchanger must be dug or vibrated into the soil.

Heat stores using solid–liquid phase changes are the most interesting from a heat transfer view point. There is basically a choice between encapsulating the material, so that the heat transfer fluid is always in contact with a solid, or having a bulk container of the material and immersing a heat exchanger in it. For most applications, encapsulation is cheaper; however it requires specialized automatic production plant. Table 11.1 gives a general comparison of the merits and disadvantages of these two techniques.

Table 11.1   *Heat exchanger configurations for phase-change material (PCM) based stores*

Generic type	Principle	Installation factors	Heat exchanger volume	Power density	Nucleation (salt hydrates)	Segregation (salt hydrates)	Cost
Encapsulated PCM	Capsules	PCM handled in factory	V. low	V. good	Poor	Fair	V. low
	Trays		Fair	Good	Fair	Good	High
	Tubes		Fair	Good	Poor	Poor	Low
Bulk PCM	Coiled tube bank	PCM handled on site	V. high	Poor	Fair	Poor	V. high

Summary

1   Sensible heat stores are generally simple, reliable, use cheap materials, but require large volumes.
2   Latent heat stores require elaborate means of heat exchange, use expensive materials which are prone to degradation, but require smaller volumes than sensible heat stores.
3   Thermochemical stores are still at the research stage and must be proved to be viable and safe in practice.

Thermal energy storage (TES) and conventional heat pumps

TES may be applied externally to the heat pump, at the 'hot' side (the load side of the condenser), or at the 'cold' side (the source side of the evaporator). Alternatively, it may be applied within the heat pump, the store forming a time-phased evaporator or condenser, depending upon the conditions under which the heat pump system is operating. These modes are illustrated in Figure 11.2.

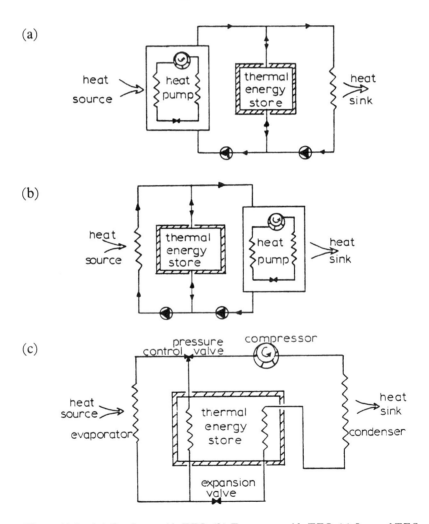

Figure 11.2   (a) Condenser-side TES. (b) Evaporator-side TES. (c) Internal TES and heat pump system.

Condenser stores

Condenser stores are probably the simplest form of TES which may be incorporated into a heat pump system. One of the most interesting applications of such a system involves the use of 'off-peak' tariff electricity. This should result in a considerable reduction in the running costs of the system, provided that any diminution in the performance of the heat pump caused by lower night-time ambient temperatures, or a requirement for higher condenser temperatures to effect the storage of the heat, is not serious. The results of an analysis of such a system, presented in Table 11.2, confirm that very low running costs are achieved when a fairly large store is incorporated.

The cost-effectiveness of such a system may be indicated by the financial pay-back period, and this will depend upon the exact specification of the system. Table 11.3 gives (from a simple analysis) the

Table 11.2 *Running costs and capacities for a typical off-peak heat pump and condenser thermal energy storage system*

Heat Pump capacity ($kW \times -1°C$)	Store capacity (kWh)	Energy consumption		Running cost (p/kWh)
		on-peak (kWh)	off-peak (kWh)	
11.6	100	0	3260	0.83
11.6	80	0	3252	0.83
11.6	60	247	2998	0.92
11.6	40	1694	1544	1.50
8.7	100	47	3217	0.85
8.7	80	42	3214	0.84
8.7	60	184	3065	0.90
8.7	40	1692	1548	1.50
5.8	100	344	2924	0.97
5.8	80	337	2925	0.96
5.8	60	267	2987	0.93
5.8	40	1324	1919	1.35
2.9	100	1561	1690	1.45
2.9	80	1555	1690	1.45
2.9	60	1554	1684	1.45
2.9	40	1492	1746	1.42

Capacities and costs relate to the house specified in Figure 11.4.

Table 11.3 *Relative pay-back period of typical, electrically-driven, heat pump systems* [1]

System description	T condenser (°C)	Store size (m³)	Pay-back period (years)
HP + resistance auxiliary	32	–	2.0
vs. electric resistance	52	–	2.3
HP + resistance auxiliary	32	–	7.7
vs. off-peak electric	42*	–	12.0
HP + oil auxiliary	32	–	5.5
vs. oil	52	–	9.2
HP (off-peak) + condenser	32	0.47†	7.7
store vs. off-peak electric	52	0.47†	16.6
HP (off-peak) + condenser	32	8.18**	7.5
store vs. off-peak electric	52	8.18**	16.0

* break-even condenser temperature = 48.7°C
† phase-change material store
** water store

pay-back periods for several representative systems. It is clear from these results that the heat output temperature is a very sensitive parameter affecting the economic feasibility of any system. Further, it also has a considerable influence upon the cost of the heat distribution system, as well as upon the practicality of the system, and the life of the heat pump.

A further surprising result is the relatively small economic benefit gained by exploiting the 'off-peak' tariff. The reason for this is that to make full use of the 'off-peak' tariff, both the heat pump and store capacity must increase to 'compress' the system operation into the 7-hour period available. The increased capital cost associated with this, off-sets the savings in running costs.

Another interesting factor illustrated in Table 11.3 is the huge volume saving possible with phase-change material stores compared with water stores. This is a consequence of the strict limitation of upper and lower storage temperatures achievable in this application. The upper temperature is limited by the output temperature available from the heat pump. Few conventional heat pumps can operate at output temperatures greater than 60°C for protracted periods and, in any case, high temperatures will depress the COP. The lower temperature is

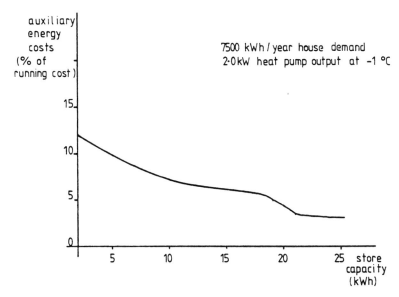

Figure 11.3   *Effect of store capacity upon auxiliary energy cost.*

limited by comfort criteria and the cost associated with the low temperature heat distribution system.

There may however be justification in developing relatively small condenser stores with heat pumps, if only because this is an effective means of matching a given size of heat pump to a wide range of building loads. In addition, the provision of a store can reduce the consumption of auxiliary energy (Figure 11.3), reduce on/off cycling under part-load conditions, and thus increase the life of the heat pump, reduce noise, and possibly reduce the cost of the heat pump by simplifying the control system.

Evaporator stores

TES applied at the evaporator side of the heat pump enables the heat pump performance to be maintained over periods when the source temperature is reduced. This is particularly important when considering air source, space heating heat pumps, which are the most common type of heat pump found in domestic situations. It is often cited that the biggest problem with conventional heat pumps used for space heating duties is that as the air temperature falls, the heat demand rises, but the

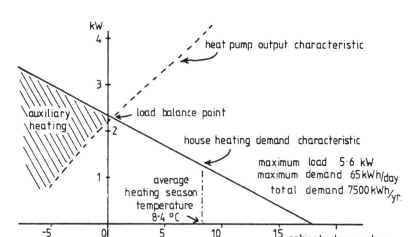

Figure 11.4   *House heating demand-load characteristics.*

heat supplied by the heat pump falls. This characteristic is responsible for the oversizing of the heat pump in most domestic applications, and the copious quantities of auxiliary heating capacity often specified. This is demonstrated graphically in Figure 11.4.

Most heat sources available to heat pumps, with the exception of the atmosphere, possess intrinsic energy storage capacities, for example ground water, or surface water. Thus, such energy sources do not vary much in temperature throughout the year, especially in temperate climates, and therefore little benefit can be gained from increasing the storage capacity artificially.

Air, as a heat source, possesses little heat storage capacity, and varies considerably in temperature on both a daily and seasonal basis. The improvement in performance which evaporator storage could achieve could be of the order of 10–50%, depending upon the system design, and would permit a smaller heat pump to supply a given heat load. Both inter-seasonal and daily storage are likely to require large sized stores.

Probably the most obvious evaporator storage medium is the water/ice phase-change system. This has the benefit that the materials are cheap and safe, although the temperature of phase-change is somewhat lower than would be considered suitable for an inter-seasonal store in a temperate climate such as is experienced in the United Kingdom. Here the average heating season temperature is approximately 7°C, and temperatures of 0°C and lower are rare (forming approximately 2.5% of the heating season).

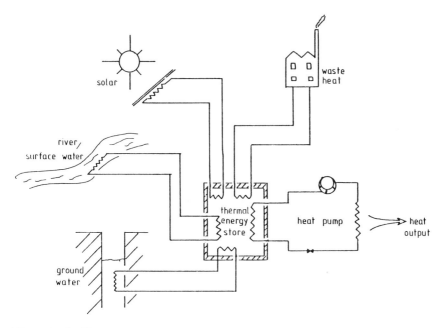

Figure 11.5   *Heat pump evaporator store regeneration – possible sources.*

A major problem on inter-seasonal storage is the very large volumes of store required (43 m³ for a 4000 kWh water/ice store). A further problem, which pertains to all evaporator stores, is that of regeneration. Figure 11.5 illustrates some of the heat sources which may be used to regenerate such a store. Domestic or industrial waste-heat, solar energy, surface or ground water, and the atmosphere are all possible sources, but care must be exercised in designing the regeneration system, as this extra cost could off-set any reduction in running costs. A major advantage of the water/ice store is that because of the high specific heat which water possesses, use may be made of the sensible heat available following regeneration. Solar ponds have been considered as evaporator stores for heat pumps at high latitudes [2], where winter ambient temperatures are low, but insolation in summer is sufficient for solar regeneration.

Internal heat storage

A third possible technique exists: this is basically a variation of the evaporator store, but with the heat pump itself providing the method of regeneration. During periods when the heat pump output is greater than the heat demand, the excess heat output is used to charge a thermal energy store. This store could form an additional refrigerant condenser within the heat pump circuit. As an alternative, the store could be charged from the liquid line leaving the condenser (see Figure 11.2c). The sensible heat held by this liquid is normally degraded as the liquid passes through the expansion valve, producing 'flash gas', unless liquid-suction heat exchange is employed to recover this heat. During cold periods, when the heat pump performance is reduced, the heat stored in the internal store may be used to augment, or replace that from the evaporator.

In addition to the 'conventional' methods of heat storage, one type of store which may suit this particular application is the 'warm ice' based store. Problems with the thermal conductivity of ice, and the temperature of phase-change, make conventional water/ice stores not entirely suitable for heat pump applications. In the USA, much of the energy storage research is directed towards air conditioning applications, and particularly at the storage of 'cooling' for summer air conditioning. Water/ice stores are not entirely suited for exactly similar reasons as those given for the heat pump applications which we have considered. As a result of this research, a 'warm ice', exhibiting high thermal conductivity, and phase-change temperatures of +4 to +10°C has been developed [3]. This is formed by evaporating a halocarbon refrigerant through a bath of water; the halocarbon clathrate formed as the heat is extracted from the water, consists of water molecules surrounding a halocarbon molecule. No heat exchanger surface is required because direct heat exchange between the refrigerant and the water is accomplished.

The control of such a system could be complex, and its effects upon the heat pump's components and its overall performance unpredictable. For these reasons it is not possible to estimate the economics of such a system, however it should be at least as viable as the externally regenerated evaporator store.

Thermal energy storage and direct-fired heat pumps

Direct-fired heat pumps fall into two categories, the engine-driven heat pump, and the absorption heat pump. The engine-driven heat pump is a thermodynamically more basic device than the electrically-driven heat pump. However, in practice, it is much more complex to design, build and maintain. The power source is a conventional internal combustion engine, fitted with heat recovery systems, the waste heat from the engine being added to the output from the heat pump.

Engine-driven heat pumps are normally high output devices, designed for commercial, industrial, and communal heating systems. They benefit from an ability to modulate their heat output, which is in any case not so dependent upon the source temperature as that of an electrically-driven unit. Further, because the heat recovered from the engine is available at a higher temperature than that at which heat pumps normally operate, the heat output temperature is normally higher than from an electrically-driven system.

Because there is, as yet, no rapid cycling of the tariffs of fossil fuels, short-term condenser-side storage of heat produces no running-cost benefits. Long-term storage at the evaporator side is also likely to be of little benefit because of the relative independence of the unit's output upon the heat source temperature; any benefit being in terms of slightly improved efficiency, without any significant reduction in size and capital cost.

The absorption heat pump possesses an intrinsic energy storage capacity as a result of the reservoirs of concentrated absorbent/working fluid which exist within the system. The storage potential of this type of system could be raised easily by increasing the size of the vessels within the heat pump, and the inventory of materials held therein. In situations where the peak-to-mean load ratio is very high, the use of an external heat store may prove desirable, in order to cope with the peak demands without requiring a high capacity (and expensive) heat pump.

Conclusions – the future

Thermal energy storage is an appropriate technology whereby the transient performance and economics of electrically-driven heat pumps can be improved. Simple pay-back calculations indicate that potential savings are likely to be low, and further studies are at present being carried out at the Cranfield Institute of Technology, under a grant from the UK Science and Engineering Research Council, with assistance from Myson Copperad Ltd.

Phase-change thermal energy storage is appropriate for heat pump applications because of the restricted operating temperature ranges, and reduced volumes required for storage. These compensate for the slightly higher operating and capital costs.

Condenser-side energy storage improves the matching of the heat pump to the load, and enables the heat pump to make use of any favourable tariffs. Any attempt to operate completely within the reduced tariff period, while reducing the running costs of the system, will increase the capital cost of the system to an unacceptable level.

Active evaporator-side heat storage can improve the seasonal COP of the heat pump, but the cost of providing regeneration of the store is likely to nullify any economic advantage of the system.

Large engine-driven heat pumps are unlikely to benefit from thermal energy storage to any significant degree. Some provision of storage, inherent in the working fluids of the systems themselves, can be achieved cheaply in the case of absorption and thermochemical heat pumps, and under certain load conditions, a small heat store may prove advantageous.

References

1   Wood R. J., O'Callaghan P. W., Probert S. D. & Buick T. R. (1982) *Storage Heat Pumps*. Paper presented at the SERC Workshop on Energy Storage Combined with Heat Pumps, Coseners House, Abingdon, Oxon., 3–4 February 1982.
2   Lund P. D. & Routti J. T. (1983) An analysis of district solar ponds with heat pump. *International Journal of Ambient Energy*, Vol. 4, No. 4, October 1983.
3   Tomlinson J. J. (1982) *Heat Pump Cool Storage in a Clathrate of Freon*. 17th. Intersociety Energy Conversion Engineering Conference, 8–13 August, 1982. IEEE no. 82 CH 1789-7.

Discussion

G. D. Braham (Electricity Council)   Swimming pools represent a specialist sector containing an inherent large thermal store which can be used in association with both the evaporator and the condenser. Using a heat pump on off-peak night-time tariff with ambient air as heat source, heat can be transferred to and stored in the pool at a COP of 2–2.2 with large quantities of heat stored for a mere 1–2 K temperature change. The same compressor and condenser can be used as a dehumidifier the following day.

T. R. Buick (Cranfield Institute of Technology) Where large volumes and large storage capacities already exist any project that can make effective use of the store at little cost is likely to be attractive. Swimming pools probably represent one of the best applications of storage with heat pumps.

G. W. Aylott (Electricity Council) In Table 11.3 an off-peak unit is compared with condenser storage using salt hydrates. A storage volume of less than 0.5 m³ was specified, whereas with a water store the volume was 8 m³, a ratio of about 16:1. Has Dr Buick any practical examples of a working salt hydrate store which can be made that much smaller than a water store. Electricity Council experience suggests that the ratio is rather less than 2:1.

T. R. Buick (Cranfield Institute of Technology) It depends upon the temperature range over which the water store can be allowed to operate. The higher the temperature range then obviously the lower the volume to be stored.

We have a phase-change store at Cranfield which uses a paraffin wax. This has a volume ratio of greater than 2.5:1, and waxes are poor in this respect compared to the performance available from salt hydrates, which are up to twice the storage density of waxes. (See Figure 11.1.)

R. R. Cohen (Cranfield Institute of Technology) With a temperature swing of 25 K, a phase-change store might be two or three times smaller than the equivalent water store. If the swing is limited to 10 K the volume reduction might be a factor of 4 or 6, depending upon the phase-change material employed.

G. W. Aylott (Electricity Council) What about the provision of hydraulic systems needed to get the heat into and out of the salt hydrate store. In our experience this takes up as much space as the salt hydrate.

R. R. Cohen (Cranfield Institute of Technology) Typically a phase-change material occupies at least 60% of the volume of a phase-change store, and in some designs over 80%.

T. R. Buick (Cranfield Institute of Technology) The phase-change store using paraffin wax under test at the moment is roughly half the physical volume that would have been needed if water had been used. The 2:1 ratio is considerably lower than the figures in Table 11.3, but paraffin is an organic material with a relatively low latent heat. Figure 11.1 shows that inorganic salt hydrates can have high volumetric latent heats (kJ/litre) and are commercially available as heat storage material. The effective volumetric heat storage does indeed depend upon the way the storage material is packed and quite often the amount of water needed for heat exchange will reduce the real values considerably beneath those listed in Table 11.3.

Dr M. A. Bell (Plymouth Polytechnic)   The authors' presentation showed a graph, based on Table 11.2, of capital cost, store capacity and energy consumption with a line of constant running costs, although this involved a variation in capital cost. Is there an optimum heat pump size and store size based on cost, extended over the life-time of the system?

T. R. Buick (Cranfield Institute of Technology)   As part of present research being carried out under a grant from the Science and Engineering Research Council, it is intended to try to form an optimum system for a given heating load which will enable reduction of the overall cost (not just running cost) per kilowatt hour. The system cost, the size of the store and the size of the heat pump will be optimized so that, overall, the system has a much reduced pay-pack period. The problem at present exists in trying to formulate how the performance of the heat pump is affected by the store, because the store itself affects the operation of the heat pump – heat must be supplied to the store at a higher temperature than would be necessary just to put the heat directly to the house.

There are also several other effects which mean that the manufacturer's performance curves cannot be used directly to determine the overall performance, the COP and the heat output of a given heat pump on a seasonal basis. Our research is to try to perform a computer optimization to establish that a certain size of heat pump and a certain size of store are the best combination for a given heat load under typical British climatic conditions.

R. R. Cohen (Cranfield Institute of Technology)   In the domestic sector there is a limited range of heat pumps. Adding a heat store to a particular heat pump makes it more versatile in the range of houses it can heat.

Dr J. Paul (Sabroe & Co.)   Could the authors give a short description of the production and application of 'warm ice' clathrates?

T. R. Buick (Cranfield Institute of Technology)   This work has been
developed in the USA (see reference 3). If a freon is evaporated
through a tank of water, as the freon evaporates it cools the water
down and instead of forming pure ice, the tendency is to form a
clathrate which is basically a very large halocarbon molecule which has
surrounding it a large number of water molecules. Depending on the
particular refrigerant being evaporated, different forms of these
clathrates are produced which have different melting point
temperatures. The main reason for this research in the USA is for air
conditioning cool storage because there is no need to go down to the
low temperatue of pure ice, that is 0°C, or just below. With higher
temperature cool storage the chiller performance is much higher and
because the thermal conductivity of clathrates is much higher, heat
exchanger performance is higher and much less heat transfer surface
area is needed than with present ice storage.

It is a closed tank, direct heat exchange system using the freon itself
and that is why it was particularly listed as an internal heat storage
system because it can be built within the heat pump. Obviously there
are other technical difficulties which may prevent its use.

Dr J. Paul (Sabroe & Co.)   It is common on the continent for big
buildings to have sprinkler systems for fire extinguishing and the
supply water storage tanks placed on top of the buildings are in some
cases used also as heat storage, either for cold or for hot water. For hot
water it may cause problems with the structure because of water
vapour but I know of one specific case where this is done and it
provides a considerable heat storage capacity more or less free of
charge.

R. R. Cohen (Cranfield Institute of Technology)   Unfortunately in the
UK safety regulations do not allow water reserved for fire quenching
to be utilized for any other purpose, although the possibility has been
suggested.

12 The market for heat pumps: assessment and development

T. McDonnell

Summary

The growing preoccupation with heat pumps by the Building Service Industry can be traced back to the oil crisis of the 1970s and the accelerated rate of activity in 1980 coincided with the gas supply shortages of the same period.

However, as the supplies of finite fossil fuels have become more stable in the past 2 years, so this has had a steadying influence in the industry at large, which in turn has been reflected in the approach to heat pumps. The typical application for a heat pump can now be more readily codified than had previously been the case.

This chapter sets out to identify the main market area that presently exists in the UK and examines some of the factors that contribute to this state of affairs. Looking to the future, there are two main development areas that will be explored: (1) the increasingly new and original heat pump applications that engineers are employing in their design schemes; and (2) the growing need for certification and standards.

In many respects these two developments are evolving with opposing forces – if you like the concepts of innovation contrasting with stability – but they can and should be reconciled for their mutual benefit and to the long-term advantage of the industry at large.

Introduction

Historians claim that in order to appreciate the present we must first examine the past. In other words, what have been the preceding events that have led to the situation today whereby heat pumps feature so widely in designers' considerations?

This retrospective notion is well suited to members of the construction industry, who march bravely forward, while firmly looking backwards. Empirical practice counts for so much.

So what are some of the historical factors? In the beginning W. Thompson, who later became Lord Kelvin, first observed the heat pump principle in the mid-nineteenth century.

Since that discovery little development has taken place for over a century, although there have been a number of notable but isolated heat pump installations in the more recent past, for example the Royal Festival Hall in 1951, Nuffield College Oxford in 1960, and so on. All of which strongly contrasts with current practice in Europe, North America and Japan, where heat pumps are installed in buildings, numbering thousands per annum. What then has caused this dramatic expansion in heat pump activity? Obviously there have been many contributing factors, but surely the most significant item has been, and still is, the motivational push given by the famous oil scares of 1973 and 1979.

Oil price increases quadrupled between 1973 and 1974, and prices rose continuously at a faster rate than inflation throughout the seventies. Although the price of crude actually fell for the first time in 1983, the general trend has been set in the public's consciousness of an ever upwards spiral. In addition, the political uncertainty and doubts regarding supplies and reserves have hastened the exodus from oil – in short people do not wish to be dependent on oil it would seem, to any degree.

The fuel supply situation in this country received another enormous shock when the Gas Corporation decided in 1980 to restrict new supplies for housing, industrial and commercial market sectors. It was during this period designers were forced into evaluating alternative heating supplies. It allowed engineers to examine the advantages that electrically driven heat pumps could offer them and their clients. The heat pump had indeed much to offer: it was a fully automatic heating device, it did not require ancillary storage areas, it did not explode and it did not create any pollution due to the by-products of combustion (it did not require a flue).

Figure 12.1 summarizes the situation, and it suggests that apart from the inherent features that heat pumps are able to offer the industry, it was perhaps the absence of alternative energy supplies that caused the initial and quite tremendous surge in growth of heat pump applications.

Indeed, in many respects the heat pump appeared just a little too good to be true, and to judge by some of the national and trade press

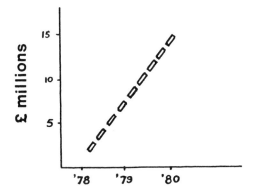

Figure 12.1   *Growth of heat pump sales 1978–1980.*

comments at the time, the heat pump was about to revolutionize the heating industry.

In addition, there was ill-informed comment from all quarters in the industry hinting that the heat pump could somehow offer something for nothing, for example COPs without qualification were being quoted as anywhere between 3 and 5 which of course totally falsifies the attractions of running costs for a heat pump. With such an inflated reputation it is perhaps surprising that so many schemes have gone ahead with such relative ease.

During this period, between 1979 and 1982, a large percentage of heat pumps were installed, primarily for space heating purposes, and the air conditioning was almost an optional bonus, although it is difficult to justify a heat pump for heating only because of the disproportionate capital costs.

Paradoxically, now that gas is again relatively plentiful, it has helped to create a more stable environment in which to evaluate and select a heat pump. It is more likely that a heat pump selected on merit will be more suitable for its application than a heat pump that is to be installed in a building – simply because there is no alternative. Therefore, when cost comparisons are made for a heating only system, gas-fired equipment will invariably appear more attractive than a heat pump on a first-cost basis.

As an example of the difference in capital costs for heating only systems, this can be demonstrated by looking at a domestic heating example.

(a)   5 kW output heat pump: £1,340
(b)   5 kW output boiler: £ 250

   Difference   £1,090

Apart from the extra capital involved in buying a heat pump, there are extras such as the additional size of system radiators to compensate for the lower flow water temperature associated with heat pumps.

However, in mainland Europe the domestic heat pump market has always been more buoyant than in the UK, particularly in France and Germany. This situation has been assisted by government agencies in both countries offering tax incentives to owners of heat pumps as an inducement to move away from oil-consuming equipment.

The withdrawal of tax concessions in France, together with the go ahead of the Siberian gas supply installation, has contributed to a crash in this market around 1982 – and at present there are no signs of an obvious recovery in sales and installations.

The UK market presently tends to be dominated by air to air machines that account for approximately 80% of the market [1]. This dominance is probably due to two main factors:

1   Ambient air is the most commonly available heat sink, and air conditioning systems rely on a ducted air supply.
2   There is a great deal of North American experience already in existence in the development and application of this type of machine which allows the rapidly burgeoning home market to proceed with this vicariously won experience.

This contrasts with the practice in France and Germany where the air to water machine has been traditionally favoured. Another contrasting approach is that the heat pump is viewed exclusively as a heating only device. However, with the decline in their air to water bivalent heating market, we may possibly see this mix alter in the future.

The drive source for all heat pumps throughout Europe is essentially the electrically-driven motor to power a compressor. There are a number of notable exceptions of gas-driven heat pumps primarily in West Germany and the Netherlands – however taken overall the number of such machines is numerically totally insignificant [2].

Looking more closely at the UK market, what are some of the characteristics that help identify it as it exists today?

The heat pump really comes into a unique and dominant position in the air conditioning sector of the industry, because it can of course provide comfort cooling to the occupied spaces during the summer season and space heating in the winter served from the same package.

Although a heat pump is more expensive than an equivalent air conditioner (by 5–15%), it can show considerable overall savings on the total project by greatly reducing or omitting a separate heating system.

Under ideal circumstances it can save the costs of a boiler (or boilers) heating pipework, a heating coil or radiators. It can also save on the labour cost that would otherwise be associated with installing these items – and it can save on maintenance costs again because there are fewer items of plant and pipework to service and maintain.

About 70% of all heat pumps installed are 10 kW, or smaller [3] and so the market appears to comprise a very large number of small installations. This view is endorsed by my own company's experience where typical installation would comprise a small single package A-A or split A-A system connected with direct expansion pipework serving a high street bank, restaurant, small office suite, and so on.

Why should there be such a proliferation of smaller units? Possibly this is due to the structure of the market. New building construction is at an all time low and refurbishment of office accommodation is correspondingly high. Packaged heat pumps are compact and can be easily taken on to site through existing door-way openings. They can be located on the roof because of their relatively lightweight construction. A number of smaller units allows zoning to be introduced into the office accommodation – so that one floor can be refurbished at a time. Or one heat pump can provide cooling to a south facing orientation, while a second heat pump can simultaneously supply heating to a north facing office suite.

Smaller decentralized plant can offer better localized control with a quicker response to changing internal loads compared to large central plant.

There is also a greater degree of standby, inherently contained in a greater number of smaller units than one item of large plant performing the same duty.

If only one floor of a multi-occupational office block is occupied, and therefore needs heating or cooling, then the plant for just that floor needs to operate – rather than run the central plant-serving the whole building – just to satisfy one specific area.

Typical heat pump installations can be categorized as commercial offices, high street stores, shops and supermarkets, public buildings, restaurants, clubs and licensed premises.

In an archetypal store, the sales area would be expected to be served by a number of air to air heat pumps for good air distribution and standby. The heat pumps would then provide heating during the winter season and comfort cooling in the summer. Experience has shown that during the winter season a daily pattern of heating and cooling occurs. That is, the heat pumps operate in a heating mode to preheat the store before and during operating hours, and as the morning progresses and the lighting and occupancy load increases (the heat gains balance the heat loss and the heat pump will switch off), but still continue to circulate air to the sales areas. If the heat gains continue to rise, and this is invariably the case, then the cooling cycle would be brought on to maintain the desired comfort level within the store.

Most applications follow this pattern in so far as that they are either heating or cooling, that is they are not expected to do both simultaneously.

Summarizing this assessment of the heat pump market in the UK today:

1   Heat pumps can offer the designer and his client both attractive capital and operating costs when there is an established need for both comfort cooling and space heating in the building.
2   The case for heating only systems is made difficult because of relatively high capital costs of a heat pump compared to boiler systems.
3   Heat pumps are exclusively driven by an electric motor connected to a compressor.
4   The market is dominated by air to air machines, rated at 10 kW or below.
5   Total sales have grown from an insignificant base in 1979 (less than £1m) to its present level of around £25m per annum.

As a final observation on this section, the technology we now almost take for granted was not available to the industry as short as 8 years ago, which suggests a high level of market acceptance.

But what of future developments? If the shortage of fossil fuels (real or imaginary) contributed to developing and establishing a heat pump market, in a cause and effect way, could the reverse situation also occur? That is, now that gas and oil supplies are again plentiful does this mean that there will now be a corresponding decline in heat pump activity?

The short answer seems to be an emphatic 'no'. Figure 12.1

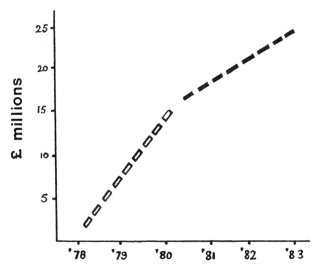

Figure 12.2   *Growth of heat pump sales 1978–1983.*

illustrated the rapid growth of heat pumps during the 1979 oil crisis and gas shortage. If this period is extended (Figure 12.2), then it can be shown that the rapid growth of heat pumps has continued to rise even though all fossil fuels are freely available. The growth rate is not as high as the initial early phase which was exceptional, but it is considerably higher than other cost yardsticks such as GNP or inflation rates.

Could it be that the relative merits of the heat pump are now more fully understood and appreciated by a wider audience and they are more able to account for themselves on practical merits so that they have to a large degree become independent of variations of fossil fuel supplies. Indeed, some of the design solutions they can offer are so unique – such as the dehumidifying applications in swimming pool halls – that there is no comparable alternative. Not only do they offer a physically more compact design solution, but they are less expensive to install and operate than were earlier plenum ventilation systems.

There are increasing indications that the trend is growing whereby heat pumps are being more widely applied in original and innovative schemes, offering scope for greater savings on both installation and running costs [4].

In general terms, this means that wherever a surplus of heat occurs within a building where simultaneously there is a requirement for a heat input, then there will always be an opportunity to literally pump the heat from one part of the building to another (before exhausting the heat if need be).

Not only can heat be redistributed around various parts of a given building, but the heat can also be transferred to and from various media. Heat removed from an air circulating system can be transferred into yet another separate air system or water system, and of course vice versa. Only the heat pump is capable of exchanging heats and media in this way and at the same time increasing the heat content in the system.

The diversity of application is only restricted by the imagination and ability of the designer.

What other developments can we expect or hope to see in the future? What for instance can we expect to see in the important area of standards and certification? Is there a need for such standards?

We surely need some sort of guideline so that we can compare different manufacturers' equipment in the following categories:

1   To ensure equipment meets safety laws.
2   To establish a common standard of rating a heat pump, so that a viable assessment can be made between alternatives for design selection.
3   Draw up an agreed set of terms and definitions describing heat pump operation and component identification.

To take one example of the confusion that still exists in the day to day dealings with heat pumps, let us consider COP. It is obviously important that this value be clearly understood by the designer if he is to accurately assess the viability of a heat pump from an operational cost point of view.

However, as Figure 12.3 [5] demonstrates, the COP for any heat pump is a constantly varying ratio (unlike a fossil-fuel boiler where the output is relatively constant regardless of outdoor or water flow temperatures).

The curves in Figure 12.3 show the dramatic difference between the COP of an air to water heat pump and the COP of resistance heat, especially at higher outdoor air temperatures. At −10°C and with a hot water supply temperature of 50°C, a typical heat pump has a COP of 1.4 which means that for each unit of power input, the heat pump supplies 1.4 units of heat output. For a hot water supply temperture of 40°C and an outdoor air temperature of 15°C, the same heat pump supplies 4.0 units of heat output for every unit of power input. Its COP has risen to 4.0.

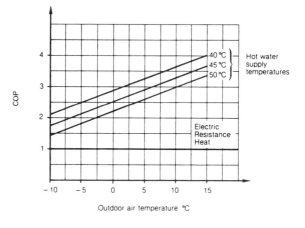

For resistance heat of any type, the COP can never exceed 1, i.e. one unit of input never produces more than one unit of heat output.   (C A R R I E R)

Figure 12.3   *Relationship of hot water and ambient temperatures on COP.*

So we can see that the range of COPs for the same heat pump can vary from 1.4 to 4.0.

Which value should the designer apply to his calculations? If he is considering two or three selections from competing suppliers of heat pumps, can he be sure that they are all quoting on a common basis? Would it be more helpful if all such quotations were in accordance with a common set of conditions? If you feel that this would be beneficial what standard should be adopted?

If you are of the opinion that such a set of standards covering performance ratings, definitions and safety laws would be beneficial to the industry at large, then you may or may not know that such an interim set of standards is already in existence. There are two sets of standards [6], one covering heat pumps up to 15 kW output and the second covering the range 12–70 kW. They have been produced by the British Air Conditioning Advisory Board (BACAB), and comprise representatives from HEVAC, the Refrigeration Industry, HVCA and the Electrical Supplies Industry. The BSI have expressed an interest in using these interim documents as the possible basis of producing their own standards.

If you are of the view that such a set of standards is important, in encouraging a better understanding of terms and definitions relating to all heat pumps, together with performance ratings including a reference to safety laws, then these standards will only occur when free market forces dictate. In other words, if you feel strongly on this topic and if you make your opinions felt in sufficient numbers then this will help force the market into collectively accepting and actively entering into a system of common standards.

Hopefully we can use conferences, such as those run by the Construction Industry Conference Centre Limited, to raise the industry's collective consciousness to consider the longer-term implications of testing and certification:

1 To develop an independent test facility to measure and assess performance ratings of heat pumps with agreed test criteria using a common datum reference.
2 To be in a better position to influence and react to future legislation such as the Energy Acts.
3 To enable the UK to speak with a single agreed voice at International Standards Organisation (ISO) and European Heat Pump Standard (SEN) meetings that are presently being set up in Brussels.

If these issues are not faced up to now by the UK Industry, we may find that our contributions to the proposed SEN and ISO standards will not fully reflect the views and experience that we have collectively acquired, and consequently both these standards when finally drawn up could well be inconsistent with our understanding and practices in the UK.

Conclusion

The initial interest in heat pumps was created mainly by the fossil-fuel shortage of the 1970s. However, in today's market the heat pump is firmly established on merit – in so far as it can offer the designer an energy-efficient system coupled with competitive installation costs.

The value of the market has increased from a low base in 1978 to around £25–30 million today; all indications are that this market share is increasing at a faster-than-average rate.

Although standards covering performance ratings are being drawn up, this important practice seems to have a low priority in most sections of the industry within the UK which contrasts starkly with our European counterparts.

References

1 *Research among installers and users of heat pumps*. Electricity Council, July 1982.
2 *Heat Pumps*. Journal CIBS, December 1983.
3 *Project profile heat pumps*. BSRIA, February 1984.
4 *Heat pump progress*. Journal CIBS, June 1982.
5 *Electric heat pumps in HU systems*. Carrier, 1982.
6 Interim standard 15 kW output, BACAB, January 1983.
   Interim standard 12–70 kW output, BACAB, January 1983.

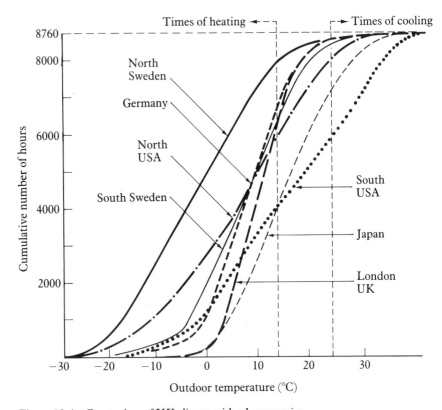

Figure 12.4 *Comparison of UK climate with other countries.*

Discussion

J. Mundy (Cybus Energy Services Ltd) Many people may get the idea that in order to heat a warehouse using a heat pump it must be air conditioned and such a view would be quite wrong. We ourselves must go out and create the market, and if we consider that everything has to be air conditioned this does not spell much of a market for heat pumps. There are many buildings, warehouses and storage facilities that will require heating perhaps overnight when the outside ambient drops (and off-peak electric tariffs are available) and I would suggest a heat pump on overnight tariff is both economic and a good application. When the system switches off at 0700 hours with a temperature perhaps 2 K higher than would normally be required, further heating will not be required for another 4 or 5 hours. This latter effect is even more pronounced in factories where people and equipment start 'giving out heat' around 0700–0800 hours. This heating only application should not be forgotten.

There are quite a number of heat pumps used for extract air recuperation and quite a lot of buildings require 100% fresh air which can all be returned through an exhaust extract heat pump.

T. McDonnell (Carrier UK Ltd)   In this chapter I attempted to report as objectively as possible the facts as I see them, and all the facts relating to the market place indicate it is a market totally dominated by air conditioning. I have observed delegates at this conference doing mental somersaults trying to justify a heat pump heating market when basically it does not exist. The capital costs really are so high initially compared to any other form of heating device that even though you do get theoretical operating savings applying a heat pump against other types of fossil-fuel systems the savings are really just outweighed by the initial capital cost. Also, if we try and artificially force the heating market, or simply force heat pumps into a heating market the result will be a repeat of the experience in mainland Europe reported when Dr Paul (Chapter 7) expanded on the collapse of that particular market. In theory it looks attractive but in practice it can be demonstrated that it is not financially viable.

Dr W. M. Currie (Energy Technology Support Unit)   At the conference 'Heat Pumps in Buildings'* Jim Leary got a lot of stick from delegates because the Electricity Council was involved in a local authority office building which had been air conditioned using heat pumps. Delegates attacked him and said 'Why on earth did you design a building that needed air conditioning when building and site constraints would have enabled avoidance of air conditioning?' It seemed to me that Mr

* Sherratt, A. F. C. (1984) *Heat Pumps for Buildings* Hutchinson, London.

McDonnell was arguing that rather than condemning that kind of thing as sinful it should be recognized that it is being done in other countries, including Japan, which is not a particularly tropical country, and that perhaps we should revise our attitudes with growing standards of living and begin to think generally in terms of air conditioned buildings. Can we incorporate heat pump technology cost-effectively, and should we get away from our puritanical historical attitude towards air conditioning?

T. McDonnell (Carrier UK Ltd)   In the UK there is still a prevailing attitude of scepticism towards the need for air conditioning in commercial buildings. However, the evidence suggests that the air conditioning industry in this country is absolutely enormous. The average man in the street may say that air conditioning is unnecessary because of our short summers, he will reason that it is a luxury, and just wastes natural resources. In fact, in the modern built environment it is not just climate that justifies the need for air conditioning, it is other factors such as heat gains from lighting, occupants, information technology equipment, etc. No matter how hard you try you cannot dismiss them. We are prepared to pay money to provide space heating, but somehow we have a puritanical approach when it comes to space cooling.

J. W. Kew (Building Services Research & Information Association)   In presentation Mr McDonnell quoted BSRIA as saying that 58% of all heat pump sales are for air conditioning purposes and compared that with the Electricity Council's figure of 90% and his own organization's estimate of 95%.

BSRIA's February 1984 *Product Profile* states that 90% of all heat pump sales in the UK are used for air conditioning purposes.

Mr McDonnell suggests that the market for air conditioning is growing rapidly, on what source does he base this view?

T. McDonnell (Carrier UK Ltd)   My own company during its last complete year of operation grew by one-third. Considering the tough state of the economy how is it that people are spending so much more money on the so-called unnecessary luxury of air conditioning? Other air conditioning companies have also enjoyed significant rates of expansion, such as Myson and Airedale. Their heat pump business also expanded. These claimed growth rates seem to fit in with the view that most certainly over the last year or two the air conditioning market has grown substantially.

J. W. Kew (BSRIA)   I think Jim Webster who forms the statistics at BSRIA would agree that there is a shift from central station air conditioning systems into packaged equipment. The growths in business quoted may just be a reflection of that fact. I do not believe that the total air conditioning market is growing in the fairly substantial leaps suggested.

T. McDonnell (Carrier UK Ltd)   The growth in business relates to turnover in all aspects. My company is the biggest distributor in Europe so we are talking about a significant business activity. We manufacture and market everything from window units through to very large centrifugal and absorption chillers. Heat pumps represent a fairly small but growing part of our total turnover.

G. D. Braham (Electricity Council)   For small packaged equipment there does seem to have been a shift to a higher proportion of air conditioning only systems. BSRIA's figures indicate a consistent growth in the number of 'Versatemp' type units sold, however in the last year or two cooling only units have had a significant proportion of the market share for small packaged units (cooling capacity 4–7 kW). Above that size heat pump packages have maintained tremendous market interest and this is reflected in the responses received to the Electricity Council's advertising.

J. Phelan (Property Services Agency)   I support Mr McDonnell's claim that the air conditioning market is increasing. In the south-east especially, where air conditioning in privately developed rented office accommodation attracts an enhanced rental which often more than offsets the cost of providing air conditioning, architects design light-weight sealed, tinted glass-fronted buildings which require constant artificial lighting and air conditioning. Many recent developments in Croydon reflect this trend.

In 1974 I looked at the viability of air to air heat pumps for domestic heating in competition with oil/gas-fired boilers. It was obvious that with the then current technology (and nothing has greatly changed) unless the cost of electricity could be freed from the cost of oil/gas by some form of subsidy, it was unlikely that domestic heat pumps would become viable in the UK.

G. W. Aylott (Electricity Council)   The cost of electricity has very little to do with the viability of domestic heat pumps. Very extensive trials have been carried out and are continuing. The price of electricity influences the demand for a whole range of heating products, because it is an energy which is usable in a number of different ways. When those products are cheap or relatively cheap, people buy them as evidenced by the increased sales of storage heaters in the last 3 or 4 years, since the price of gas has increased so much relative to electricity. The reason domestic heat pump sales have not gone up similarly is because they are (a) very expensive and (b) not at all reliable.

Dr R. G. H. Watson (Building Research Establishment) Condensation is becoming a major problem in housing, particularly in the low-cost end of the spectrum where people cannot afford a lot of money to heat their houses. We have heard that air conditioning is a major market area for heat pumps. Can we look forward to either a development of heat pumps or a development in the way in which they are inserted into the building to help the local authorities overcome this problem of condensation in homes?

Dr D. J. Martin (Energy Technology Support Unit) As far as dehumidification is concerned there probably is not a great deal that needs to be done or indeed can be done other than reduce the capital cost. Cost is the biggest drawback and future developments must centre on cost reduction. This opens up the question of where the capital cost reductions can be achieved and one means would be by volume production. Larger numbers of units being manufactured would in theory bring down the price per unit but clearly until the market develops this will not be possible.

T. McDonnell (Carrier UK Ltd) There is available on the market today a single package air to air heat pump suitable for indoor installation where the outdoor air supply can be ducted to and from the coil. Dr Watson has identified in the condensation problem a rather specific market which possibly no other form of heating device can satisfy.

Dr W. M. Currie (Energy Technology Support Unit) The chief problem that local authorities claim is cash. I have heard some passionate expositions by local authority personnel that they cannot even begin to deal with basic property maintenance, let alone consider heat pump central heating, external wall insulation and all these other things that could improve the internal environment.

Dr R. G. H. Watson (Building Research Establishment) One problem local authorities face is the cost of renovating the damage done by condensation. If this could be weighed against the cost of a better solution, there might be a rather more rational choice.

Dr J. Paul (Sabroe & Co.)   Initially I was amazed at the difference between Germany and the UK portrayed by Mr McDonnell, but I now realize that what he described as air conditioning heat pumps in Germany would not count as heat pumps at all because they are units bought mainly for cooling purposes and have an option of heating. My negative or depressing comments earlier (Chapter 7) about the German heat pump market referred only to heat only heat pumps. The air conditioning market where the cooling system is used also for heating is more or less the same in Germany as in the UK scene indicated by Mr McDonnell. Everyone is checking whether they can make use of the condenser heat. It is the German domestic heat only market, which was much bigger in percentage of market share, that has shrunk.

J. W. Kew (Building Services Research & Information Association)   Dr Currie mentioned Japanese use of reversible heat pumps for cooling and heating. A curve of outdoor temperatures against cumulative hours of the year helps to explain the reason for this.

If the need for cooling is taken solely as a function of outside air temperatures above 25°C, the hours per year that the temperature exceeds 25°C are as follows: Japan, 1400 hours/year; South USA, 2700 hours/year; North USA, 800 hours/year. By comparison the figure for the UK is 80 hours/year.

This I believe explains the popularity of reversible heat pumps in Japan and the USA. In the UK we cannot justify cooling on the grounds of temperature alone.

I agreed with Mr McDonnell that there are other good reasons for providing air conditioning. As well as the points he made, there is an additional item – the commercial reason for air conditioning. This provides an answer to Mr Phelan who noted 'the profileration of air conditioning in Croydon'.

Developers can charge a rent premium for air conditioned offices in London: 25% central London and 40% fringe City of London. Both of these rent premiums for air conditioning worked out to £50/m² according to a Hillier Parker survey in July 1982.[*]

---

[*] 'The effect of air conditioning on office rents'. Research Report No. 5, July 1982. *Investors Chronicle*, Hillier Parker Rent Index.

Compare £50/m² with the £100/m² or so for capital costs of air conditioning given by Leary & Austin (Chapter 8, Table 8.6) and I think you can see that landlords in London have a powerful financial inducement. The additional rent income can pay back the extra cost of air conditioning compared with heating in just over a year! This premium only applies, according to Hillier Parker, in London.

It is believed that a high percentage (around 80%) of all offices are in developer/landlord ownership.

Figure 12.4 (see p. 240) shows data from the government survey of the air conditioning market presented in *Business Monitor PQ 339.3** as a graph from the mid-1970s onwards together with the market trends for total HVAC. It can be seen that HVAC shows continuous modest growth. Air conditioning by contrast shows a decline since 1980. I would therefore contest Mr McDonnell's description of a 'rapidly growing market for air conditioning' – I think we all wish it were true!

Given a free hand on an unrestricted site it is possible with narrow plan carefully designed buildings to eliminate the need for air conditioning in the UK. I do not accept the argument suggested by Dr Currie that puritan ethics prevent us from air conditioning! The BRE low-energy building demonstrates my point. On Monday 11th July 1983, the hottest day in the south-east for 7 years, the outside air temperature reached 30°C:

- BRE low-energy office 24°C average, 27°C peak internal temperature
- adjacent 1960s buildings, 31°C peak internal temperature

The BRE building achieved 'air conditioned' standards for comfort without the need for cooling! If you have freedom of site conditions and normal office accommodation air conditioning can be designed out.

T. McDonnell (Carrier UK Ltd)   The government data quoted by Mr Kew is some 2–3 years old. I based my comment on the growth of the market leader and an increase in turnover of one-third cannot be hidden. There is always a time lag with statistics and the accuracy of Figure 12.4 would be questionable.

J. Phelan (Property Services Agency)   Although energy conservation and the application of heat pumps, especially in air conditioning

* '*Business Monitor PQ 339.3*'. Business Statistics Office, HMSO.

buildings, is commendable, our prime function must be to provide an acceptable working environment in which highly motivated, high technology personnel, can maximize achievement and output. A holistic approach is essential to ensure that a proper balance is reached.

T. McDonnell (Carrier UK Ltd)   The additional premium for air conditioned office space in London is surely an arrangement agreeable to both parties. However, what does give rise to concern is the way air conditioning is sometimes promoted as a sales feature in order to sell office space. Witness sign boards proclaiming 'prestigious new office to let – fully air conditioned'. Does air conditioning then become associated in the public consciousness with luxury and waste and therefore miss the whole point of its more fundamental need to provide a controlled and effective working environment. As Mr Phelan commented, we should be designing for a more stimulating, healthy environment based on the needs of the occupants – not on estate agent notions.

Properties in the City of London which are refurbished to a very high standard with good lighting, interior design and air conditioning are sublet very quickly, yet there is a surplus of properties in the suburbs of towns where air conditioning, etc. has not been installed. Is there a link here?

B. A. Bath (Rybka, Smith & Ginsler, Ontario)   Mr Kew quoted temperatures in the BRE building but not humidity levels. I suspect that, although temperatures were reasonable, humidity levels were very high and the space uncomfortable. However, I am accustomed to air conditioning.

13 Advanced heat pump technologies

D. J. Martin

Introduction

Future historians will probably view the 1980s as the decade in which the heat pump became a reality for space heating in buildings. I say this with confidence because many factors now point to the emergence of successful cost-effective designs of advanced heat pumps which will be commercialized during the next 5 or 6 years. No longer will the heat pump be portrayed as the 'jam tomorrow' of energy conservation – heat pumps will take their place as highly energy-efficient, user-friendly and consumer-acceptable energy conservation equipment.

To explain why I have so much optimism, I would like to outline some of the advanced heat pump systems being developed and to discuss the market potential and energy savings benefits associated with these developments.

The scope for heat pumps

The potential for improving energy efficiency in buildings by the use of heat pumps is enormous, and because of this there are probably few topics which excite so much discussion and interest in the energy conservation field. The Energy Technology Support Unit (ETSU) estimates indicate that the annual energy savings in the UK from the penetration of heat pumps into the major space heating markets over the next few years could be up to 0.2 Mt c.e. (Table 13.1), corresponding to energy cost savings of around £20 million per year and equipment sales in excess of £100 million. Furthermore, the UK has a long-term potential for energy savings of around 5 Mt c.e. per year from the use of heat pumps in buildings. On an international scale, the IEA has estimated that even by 1995 heat pumps could offer energy savings of 8 Mt c.e. across its member countries.

It is not surprising, therefore, that heat pumps have captured the imagination and attention of governments, manufacturers and customers alike. However, progress in the major heating markets has so far been very slow. This is because of the high first cost of heat pumps and persistent doubts over their reliability and performance. As with all other energy saving devices, the heat pump must be able to compete with conventional heat generators and be compatible with existing heating systems: up to now the increased fuel efficiency of the heat pump has rarely provided adequate compensation for its higher capital cost and increased complexity compared with established technologies.

Table 13.1   *Energy savings from the use of heat pumps in the UK*

Space heating market	*Time scale*	*Energy savings kt c.e./year*
Current specialized applications (swimming pools; retail trade; oil boiler replacements)	Current	50–60
Early market applications (mostly in non-domestic buildings)	During the next 4–5 years	100–200
Major market applications (non-domestic and domestic buildings)	During the next 10–15 years	Up to 1000
	Long-term potential	Up to 5000

Nevertheless, there are many forces at work encouraging progress in heat pump technology. The objectives of advanced heat pump R&D are to improve their cost-effectiveness and reliability, with the aim of giving them more direct appeal to customers in mass market applications. If successful, this should enable heat pumps to achieve their market potential.

Heat pump R&D

The thrust of most advanced heat pump developments is for products aimed at the domestic and light commercial buildings space heating markets. The reasons for this are not hard to find. These markets offer very large volume numbers of sales and have a reasonably uniform set of technical and commercial requirements. Successful heat pump designs developed for these markets would provide longer production runs and present fewer specialized installation and maintenance problems than heat pumps for industrial process heating or large commercial buildings applications.

There are three principal technical categories of heat pump R&D work. These are:

1   Improving components in existing heat pump designs.
2   Extending the operating range of heat pump cycles.
3   Developing new systems and operating cycles.

All three categories are the subject of extensive research and development, with work being carried out by equipment manufacturers, research organizations and the fuel supply industries. This chapter describes examples of projects in these categories and for the most part will discuss projects aimed at space heating in buildings.

Improving components

Most current designs of heat pumps are based on the vapour compression cycle. This is a fairly well-known technology which has been successfully employed in refrigeration and air conditioning. However, there is still scope for component improvement where it is applied to heating only duties. Among the component improvements being carried out are:

* non-azeotropic mixture refrigerants
* capacity control of electric motor-driven heat pumps

Non-azeotropes

One way of improving the performance of a vapour compression heat pump is to increase the efficiency of heat transfer at the evaporator and the condenser. The temperature of a conventional refrigerant remains virtually constant through the evaporation or condensation stages, while the temperature of the fluid outside the refrigerant coil generally varies. This is clearly illustrated in the well-known Carnot cycle TS diagram (Figure 13.1). However, heat transfer would be more efficient

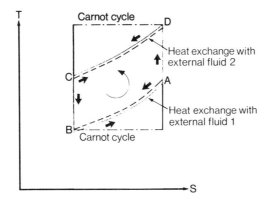

Figure 13.1 *Theoretical reversible cycles.*

if the refrigerant temperature varied by following the temperature of the outside fluid, as illustrated in the Lorentz cycle T S diagram (also on Figure 13.1). It should be possible to achieve this by using a mixture of refrigerants which condenses and evaporates over a range of temperatures: a non-azeotropic mixture.

Non-azeotropic mixtures make it possible to improve the COP or to increase the thermal capacity of the heat pump. These potential benefits have led to experimental work on the selection and testing of mixed refrigerants, most notably in France and in the United States. Manufacturers' interests in this topic are illustrated by Dupont, who have given details of an experimental mixture of Freon 13B1 and R152a, described as offering a performance improvement over R22. Figure 13.2 shows

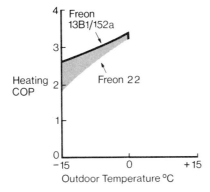

Figure 13.2 *Laboratory test data for heating COP versus outdoor temperature.*

some laboratory data published by Dupont and indicating the enhanced performance obtained with this non-azeotropic mixture [1].

In France, research at Institut Français du Pétrole (IFP) has been aimed at selecting a suitable mixture for use in a standard domestic heat pump. This approach involves some changes to the refrigerant flow control valve because of the mixture properties at this point of the cycle, but no other major changes to the heat pump hardware. Both air to water units and water to water units have been tested and increases in thermal output of around 20% have been obtained when compared with the same units using R22 [2].

Capacity control

Conventional electric heat pumps for residential and light commercial applications achieve capacity control by intermittent operation: they cycle on and off at a frequency necessary to match approximately the heating load. Under part-load conditions, which occur most often (dependent on the relative sizing of the heat pump), this on/off cycling has an adverse effect on performance. Figure 13.3 illustrates how capacity control can improve the heat pump performance.

Adjustable speed motor drives are one means of achieving better capacity control. This can be done by means of inverter control. In the USA, the Electric Power Research Institute has studied various methods, and has found that square wave, voltage source inverters and pulse-width modulated voltage source inverters applied to induction motor drives are attractive technically. Continuous speed adjustment in

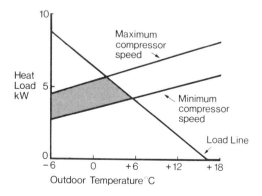

Figure 13.3 *Heat load and heat pump capacity versus outdoor temperature.*

a six-to-one range can be achieved, but the cost of the heat pump is increased by approximately £50 to £70/kW capacity [3].

Another means of capacity control is to change the compression volume in accordance with the heat load requirements. Westinghouse Corporation in the USA has developed a prototype two-capacity heat pump for domestic use which employs a dual-stroke compressor [4]. This compressor has a circular eccentrically-bored cam between the crankpin and the connecting rod. The action of the cam is such that it is driven against one of two stops depending on the direction of crankshaft rotation. The two cam positions correspond to different piston strokes determined by the cam eccentricity. This heat pump has achieved a 20% improvement in annual energy consumption compared with the most efficient two-speed units available in the USA, when operated in a space heating mode.

Extending the operating range

Extension of the operating range of vapour compression heat pumps to higher temperatures is a major objective of research aimed at broadening their applications into major new markets. In industry, they could then be used for steam raising and for high temperature heat recovery: in buildings, the need for fossil-fuelled supplementary heating would be avoided, making heat pumps more acceptable as a direct replacement for conventional heating plant.

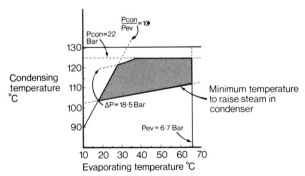

Figure 13.4   *Limits of operation of heat pump system imposed by compressor and heat transfer characteristics.*

The key to higher temperature applications lies in the stability of the refrigerant fluid and this in turn is affected by the compressor lubricant, by any other additives that may be employed and by the heat pump construction materials. In the UK, International Research & Development Co. Limited (IRD) have designed, constructed and tested a gas engine-driven high temperature heat pump [5]. The working fluid is R114, operating at a condensing temperature of 120°C and an evaporating temperature of 80°C. The prototype system operated satisfactorily with the lubricant and refrigerant combination chosen, and a COP of around 3 was obtained. This is equivalent to an overall fuel efficiency of about 140% including engine recovery. Figure 13.4 shows the operating limits of the IRD design which are imposed by compressor and heat transfer characteristics.

Developing new systems

Extensive R&D efforts are being directed into new pump systems, particularly for residential heating applications. Some exciting prospects are now beginning to emerge: these are based on absorption cycle systems, and on systems with a separate power cycle driving the heat pump cycle.

Absorption cycle systems

The absorption cycle is the best understood and theoretically the least exacting cycle. It holds promise of many advantages as listed in Figure 13.5. These include long life, low maintenance and high reliability due to the few moving parts involved. However, in spite of these advantages, exploitation of this cycle for heating applications is at present hindered by the lack of a suitable fluid pair capable of prolonged operation between the large temperature differences encountered in the cycle.

The only commercially-available absorption heat pumps suitable for heating applications use ammonia/water as the fluid pair, and are basically adaptions of existing chiller technology. These systems have so far failed to meet expectations, and much R&D is being directed towards improving their technical performance and commercial viability. A very wide range of alternative operating regimes and working fluid combination is also being studied (Table 13.2).

Worldwide, at least 25 projects involving absorption cycle heat

```
*   NO MOVING PARTS

*   LOW MAINTENANCE DEMANDS

*   QUIET OPERATION

*   COP LESS DEPENDENT ON AMBIENT
    TEMPERATURE THAN VAPOUR COMPRESSION
    HEAT PUMPS

*   EVAPORATOR SMALLER THAN IN A VAPOUR
    COMPRESSOR HEAT PUMP
```

Figure 13.5   *Features of absorption heat pumps.*

Table 13.2   *Examples of absorption heat pump working fluids investigations*

Chemical type	Pressure of refrigerant condensation	Working fluids		Investigators
		Refrigerant	Absorption	
Inorganic fluids	Low	Water	Lithium bromide	Sanyo Corp, Japan IRD, UK IGT, USA
	Low	Water	Sodium hydroxide	Cranfield Institute of Technology, UK
	High	Ammonia	Water	ASK/Ruhrgas, Germany Vaillant, Germany Happel, Germany Calor Gas, UK Arkla Corp, USA MAN, Germany
Fluorocarbons	Low	Trifluorethanol	Various	University of Essen, Germany
	Medium	R123a	ETFE	Allied Corp, USA
	Medium	R22	E181	Stiebel Eltron, Germany
	High	R152a?	Unknown	Gaz de France, France

pumps R&D are at present underway. This represents a very considerable effort, and it is worth considering in greater detail what has stimulated this degree of interest and the particular technical problems that have attracted attention.

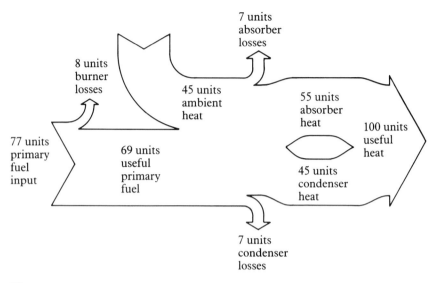

Figure 13.6   *Energy flow diagram for gas-fired absorption heat pump.*

Absorption refrigeration systems have been available for many years, and the equipment has been reasonably successful in many applications. The absorption cycle can in theory supply heat for space and water heating at much higher efficiencies than conventional heat generators and this is illustrated in Figures 13.6 and 13.7 which show the energy flow diagrams for a gas-fired absorption cycle heat pump and for a conventional gas boiler. The overall fuel efficiency of the heat pump is 130% compared with the boiler efficiency of 80%.

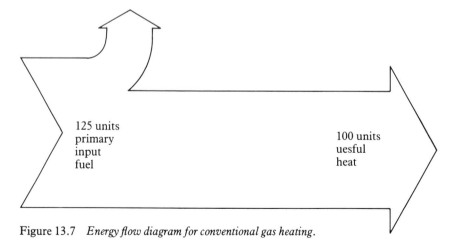

Figure 13.7   *Energy flow diagram for conventional gas heating.*

However, the transition to a device aimed at supplying heating rather than cooling has not been an easy one. Several technical problems have arisen, specific to the particular fluids used in the cycle, and these problems have proved difficult to overcome.

The problems for the ammonia/water fluid pair include:

- an engineering cost penalty dictated by the high pressure operation
- the need to separate vapours by rectification
- the need for containment dictated by the corrosive nature of the fluids
- the restricted temperature range imposed by the operating pressures
- the need for a low-cost solution pump

The water/lithium bromide fluid pair has also been used successfully in chiller applications, but the situation for heating applications is not much better. Problems include:

- an engineering cost penalty dictated by the low pressure operation
- the restricted temperature range due to freezing or crystallization
- the need to use additives to combat corrosion

The same sort of picture emerges when we look closely at other candidate fluid pairs which have been suggested. There always seems to be some problem or other, and to date no single refrigerant/absorbent pair is outstandingly attractive for space heating in buildings. Nevertheless, the wide range of research indicates that hope has not been given up, as evidenced in my earlier table of projects.

Furthermore, many variations on the simple absorption cycle have also been studied and among the designs being put forward are:

- resorption cycle in which the condenser/evaporator is replaced by a resorber/desorber arrangement, effectively adding a second absorption loop to the cycle. The process is under study by IFP in France and elsewhere.
- intermittent operating cycles where the generator/absorber functions and the condenser/evaporator functions are carried out in the same vessels at different times during the cycle. The Institute for Technical Thermodynamics at Aachen. West Germany is looking at this idea.
- adsorption cycles where a solid adsorbent bed reacts with a vapour refrigerant to produce a cycle similar to the liquid/liquid conventional absorption process. Among the projects in this topic, EIC

Laboratories, USA, have examined calcium chloride/methanol systems and the Technical University, Munich, is testing zeolite/water systems.

• metal hydride systems in which an exotic metal alloy combines with hydrogen to give off heat and subsequent heating removes the hydrogen; with judicious choice of metal alloy, a heat pump cycle can be operated. One organization claiming some success in this technique is Studovic Energiteknik in Sweden.

Power cycle/heat pump cycle systems

These systems are inherently more complex and potentially less reliable than absorption cycles. However, their potential heating performance is promising and some could provide a viable alternative to absorption heat pumps. A number of systems under study are shown in Table 13.3. Power cycles comprising Otto, Stirling and Brayton cycles and heat pump cycles including Rankine and Stirling cycle operation are all under development.

The major advantages of these systems are:

• high efficiency designs of prime mover are possible
• waste heat recovery from the prime mover can be used to boost the output of the heat pump to higher temperatures.
• capacity modulation of the heat output is relatively straightforward
• overall fuel efficiencies greater than 150% are achievable

Table 13.3   *Examples of power cycle/heat pump cycle systems under development*

Cycle type		Investigators	Status
Power	Heat pump		
Otto	Rankine	UK private company	Laboratory prototype
		Fitchel & Sachs, Germany	Field tests
		Gas and manufacturing industries consortium, Japan	Field tests
Otto	Linear compressor	Tectonics, USA	Prototype development
Stirling	Rankine	MTI/Gas Research Institute, USA	Prime mover units under test
Stirling	Stirling	Sunpower Inc, USA	Demonstrator unit
Brayton	Rankine	Garrett Corp/Lennox/Dunham Bush, USA	Pre-production unit in field tests

Nevertheless, as with the absorption cycle, there are several major impediments to commercialization. Most notably, the life-cycle heating performance and maintenance requirements have not been proven and initial capital costs are still too high to allow for a rapid market penetration.

Prospects and conclusions

The present generation of heat pump systems can be an effective way of saving energy and reducing fuel costs. However, their widespread use for space heating will depend not only on the realization of the technical improvements promised by all this development effort, but also on the creation of a confident and receptive market for those improved products. Technical ingenuity and engineering progress may result in improved performance or reduced capital cost, but they will not themselves be sufficient to secure progress; the 'market lead' is likely to be just as important as the 'technical push'.

The creation of the right market environment is an essential prerequisite to the widespread uptake of heat pump systems. Equipment manufacturers need to be convinced that a worthwhile market for heat pump products will emerge. Only then will they have confidence to invest in production machinery and to create the sales and servicing structure that must exist before that market can develop. Furthermore, at present the purchase price of heat pumps contains a high insurance provision to cover in-service problems; with increasing confidence and experience in a growing market it will be possible to reduce this provision, further improving cost-effectiveness to stimulate still wider application.

The high level of heat pump research is itself evidence that many companies all over the world are already willing to take the risks of R&D expenditure in anticipation of this market evolving. In many cases they are aided and supported by fuel supply companies or by government bodies. This high level of interest in advanced heat pump developments can only be beneficial.

In conclusion, then, there is an enormous potential for heat pumps to provide improved energy efficiency in our low temperature heating requirements. Wide ranging technical developments are being carried out by industry with a view to launching new and improved products on the market in the short to medium term. The path to be followed will not be an easy one, but I would suggest that the need to exploit heat pump technology through developments to meet existing and projected market requirements has a compelling logic associated with it.

Acknowledgements

This paper has made use of information acquired by ETSU through its work for the Department of Energy on its energy conservation R, D & D programmes. The opinions expressed are those of the author and do not necessarily reflect the official views of either the Department of Energy or the UKAEA.

References

1   Dupont Corporation, Wilmington, Delaware, USA. Freon Product Information RT-72.
2   Ramet C. E., Rojey A. & Durandet J. (1983) *Heat Pumps with fluid mixtures applied to space heating*. Seminar on Energy Savings in Buildings – Commission of the European Communities, November 1983.
3   Calm J. M. (1983) *Residential and commercial heat pump subprogram*. Electric Power Research Institute, Palo Alto, California, USA, March 1983.
4   Veyo S. E. & Fagan T. J. (1983) *Advanced electric heat pump dual-stroke compressor and system development*. Westinghouse Electric Corporation. Final report for US Department of Energy, December 1983 ORNL/Sub/79-2412/3.
5   Eustace V. A. & Smith S. J. (1982) *Laboratory testing and industrial applications of a high temperature gas engine-driven heat pump*. Contractors meeting on heat pumps – Commission of the European Communities EVR8077EN, April 1982.

Discussion

Dr J. Paul (Sabroe & Co.)   Although the Lorentz cycle and non-azeotropic mixtures have been available for many years there is a very simple but very practical drawback. With current technology the control system and the safety of the cycle have been dependent on the pressure, which was very easy to measure and was an indicator of temperature. With non-azeotropic mixtures which do not have a pressure/temperature relationship a great deal of electronic instrumention is needed to monitor operation of the machine.

We have used non-azeotropic mixtures in a test rig, but it is not just simply filling the heat pump with a new refrigerant and running it; current heat exchanger designs are not at all suitable for non-azeotropic mixtures, there are control problems and furthermore for service purposes, if something has evaporated or escaped from the system, what has escaped and in what quantity? Some heat pumps use say 5 t of refrigerant and if there is a loss of refrigerant it may be necessary to replace it all because you do not know what is left. There are therefore some practical drawbacks.

This chapter suggests that water vapour compression, where temperatures can be as high as 180°C, is new, in fact units are commercially available.

Dr D. J. Martin (Energy Technology Support Unit) This chapter describes a range of possibilities that are under investigation and they are by no means fully developed commercial systems. There are indeed several reservations on the practical use of non-azeotropic mixtures and some chemical companies are quite reluctant to get involved for the very reasons that Dr Paul described. One problem is the service requirements of such devices, for example, if there is a refrigerant leak or if there are changes in the composition of the refrigerant round the cycle. Other difficulties include the question of how to commission the unit. Furthermore, heat exchangers have to be optimized for the non-azeotropic mixture and design points have to be chosen very carefully. The refrigerant expansion valve and the control of that valve also have to be very carefully designed. All these problems are present with a non-azeotropic mixture.

The high temperature heat pump has been a major area of development, we know it is possible to get condensing temperatures of over 100°C and condensing temperatures of 150°C or even higher are being looked at. To understand the practical value of this development involves investigating the actual thermodynamics of the refrigerant to see whether it can give a sensible COP for the temperature lift involved. I described the IRD development which had a 120°C condensing temperature with an 80°C evaporating temperature. This gives a COP of about 3. But with a lower evaporating temperature from a waste heat stream, say 40°C, a COP of 3 would not be obtained. High temperature heat pumps should only be applied in situations where the evaporating temperature is high enough to enable a good COP to be obtained and where there is no other simpler way of using waste heat energy.

E. J. Perry (W. S. Atkins & Partners)   I completely agree with Dr
Martin regarding mixed refrigerants and also cascade systems for
using different refrigerants. These can make great improvements in
performance in large industrial plants where cooling and heating take
place over large temperature differences.

Can mixed refrigerants play a part in the heating and cooling of
buildings in view of the small temperature ranges involved? I
personally doubt whether the advantage of a mixed refrigerant system
would be applicable in this application.

Dr D. J. Martin (Energy Technology Support Unit)   The advantageous
conditions for this type of system include a large change in the sink or
the source temperature. Another factor is that the evaporation and
condensation lines should be parallel on the Mollier chart. If the
source or sink cannot meet these conditions then it is unlikely that the
performance improvement which is possible in theory can ever be
obtained. However, there may well be some applications in buildings
where the heat pump is operated not as the sole heating or cooling
device, but in conjunction with two thermal stores. The stores could
be charged and discharged over wide temperature bands. Such a
combination has not yet been investigated in any great detail, but
there could be scope for development.

Dr J. Paul (Sabroe & Co.)   No mention has been made of
development of new heat exchangers. In my opinion there is a basic
need to develop heat exchangers – especially evaporators – for using
heat sources which cannot be used with currently available
equipment; for example raw sewage water. Here is a tremendous heat
source which is linked to large accumulations of people, and ground-
water heat pumps are not possible because there is not sufficient
ground available – the City of London would present considerable
potential for using heat from raw sewage.

A scheme is successfully operating in Oslo where a 6 MW heat
output is provided using heat from the sewage of just a small part of
the city. This sewage water is cooled down close to 0°C using a
specially designed heat exchanger. On its journey to the sewage plant
it heats up again so in effect the heat is just being borrowed and added
back later.

Dr Martin said wisely that absorption heat pumps have a theoretical
COP of 1.3. Theoretical certainly – in my experience all absorption
heat pumps measured in practice had difficulties in getting close to

unity so at this level it is no more than a sophisticated gas burner. The biggest absorption heat pump was taken out after 3 years of operation because it was not working satisfactorily.

Dr D. J. Martin (Energy Technology Support Unit)   I have not come across work of note in that area. Certainly there are improvements possible and you do not have to go to the Lorentz cycle to get an improved heat exchange at the evaporator or the condenser. The sewage water heat source at Nuffield College was mentioned in Chapter 2. This installation used an intermediate water loop between the heat exchanger for the sewage and the heat exchanger for the heat pump. The system clearly produces an inefficiency in the circuit and if there was opportunity to use direct evaporation then that might well give a better performance for that particular configuration.

On the absorption heat pump, I believe that the promise is there for this technology and this is generally recognized. In practice the absorption system has not achieved the kind of performance that is really needed. Nevertheless, it is worth stressing that the Japanese have water/lithium bromide absorption equipment working satisfactorily with multi-megawatt installations for some of their industrial processes.

Index